The Mental Life of Monkeys and Apes - A Study of Ideational Behavior

by Robert M. Yerkes

CONTENTS

I

INTERESTS, OPPORTUNITY AND MATERIALS

Two strong interests come to expression in this report: the one in the study of the adaptive or ideational behavior of the monkeys and the apes; and the other in adequate and permanent provision for the thorough study of all aspects of the lives of these animals. The values of these interests and of the tasks which they have led me to undertake are so widely recognized by biologists that I need not pause to justify or define them. I shall, instead, attempt to make a contribution of fact on the score of each interest.

While recognizing that the task of prospecting for an anthropoid or primate station may in its outcome prove incomparably more important for the biological and sociological sciences and for human welfare than my experimental study of ideational behavior, I give the latter first place in this report, reserving for the concluding section an account of the situation regarding our knowledge of the monkeys, apes, and other primates, and a description of a plan and program for the thorough-going and long continued study of these organisms in a permanent station or research institute.

In 1915, a long desired opportunity came to me to devote myself undividedly to tasks which I have designated above as "prospecting" for an anthropoid station and experimenting with monkeys and apes. First of all, the interruption of my academic duties by sabbatical leave gave me free time. But in addition to this freedom for research, I needed animals and equipment. These, too, happily, were most satisfactorily provided, as I shall now describe.

When in 1913, while already myself engaged in seeking the establishment of an anthropoid station, I heard of the founding of such an institution at Orotava, Tenerife, the Canary Islands, I immediately made inquiries of the founder of the station, Doctor Max Rothmann of Berlin, concerning his plans (Rothmann, 1912).[1] As a result of our correspondence, I was invited to visit and make use of the facilities of the Orotava station and to consider with its founder the possibility of cooperative work instead of the establishing of an American station. This invitation I gratefully accepted with the expectation of spending the greater part of the year 1915 on the island of Tenerife. But the outbreak of the war rendered my plan impracticable, while at the same time destroying all reasonable ground for hope of profitable cooperation with the Germans in the study of the anthropoids. In August, 1915, Doctor Rothmann died. Presumably, the station still exists at Orotava in the interests of certain psychological and physiological research. So far as I know, there are as yet no published reports of studies made at this station. It seems from every point of view desirable that American psychologists should, without regard to this initial attempt of the Germans to provide for anthropoid research, further the establishment of a well equipped American station for the study not only of the anthropoid apes but of all of the lower primates.

In the early months of the war while I was making every effort to obtain reliable information concerning conditions in the Canary Islands, I received an urgent invitation from my friend and former student, Doctor G. V. Hamilton, to make use of his collection of animals and laboratory at Montecito, California, during my leave of absence from Harvard. This invitation I most gladly accepted, and in February, 1915, I established myself in Santa Barbara, in convenient proximity to Doctor Hamilton's private laboratory where for more than six months I was able to work uninterruptedly under nearly ideal conditions.

Doctor Hamilton without reserve placed at my disposal his entire collection of animals, laboratory, and equipment, provided innumerable conveniences for my work, and in addition, bore the entire expense of my investigation. I cannot adequately thank him for his kindness nor make satisfactory acknowledgment here of his generous aid. Thanks to his sympathetic interest and to the courtesy of the McCormick family on whose estate the laboratory was located, my work was done under wholly delightful conditions, and with assistance from Ramon Jimenez and Frank Van Den Bergh, Jr., which was invaluable. The former aided me most intelligently in the care of the animals and the construction of apparatus; and the latter, especially, was of very real service in connection with many of my experiments.

The collection of animals which Doctor Hamilton placed at my disposal consisted of ten monkeys and one orang utan. The monkeys represented either Pithecus rhesus Audebert (_Macacus rhesus_), Pithecus irus F. Cuvier (_Macacus cynomolgos_), or the hybrid of these two species (Elliot, 1913). There were two eunuchs, five males, and three females. All were thoroughly acclimated, having lived in Montecito either from birth or for several years.

The orang utan was a young specimen of Pongo pygmaeus Hoppius obtained from a San Francisco dealer in October, 1914 for my use. His age at that time, as judged by his size and the presence of milk teeth, was not more than five years. So far as I could discover, he was a perfectly normal, healthy, and active individual. On June 10, 1915, his weight was thirty-four pounds, his height thirty-two inches, and his chest girt twenty-three inches. On August 18 of the same year, the three measurements were thirty-six and one-half pounds, thirty-three inches, and twenty-five inches.

For the major portion of my experimental work, only three of the eleven animals were used. A growing male, _P. rhesus_ monkey, known as Sobke; a mature male, _P. irus_, called Skirrl; and the young orang utan, which had been named Julius. Plates I and II present these three subjects of my experiments in characteristically interesting attitudes. In plate I, figure 1, Julius appears immediately behind the laboratory seated on a rock, against a background of live oaks. This figure gives one an excellent idea of the immediate environment of the laboratory. Figure 2 of the same plate is a portrait of Julius taken in the latter part of August. By reason of the heavy growth of hair, he appeared considerably older as well as larger at this time than when the photograph for figure 1 was taken. In plate II, figure 3, Julius is shown in the woods in the attitude of reaching for a banana, while in figure 4 of the same plate he is represented as walking upright in one of the cages.

Likenesses of Sobke are presented in figures 5 and 6 of plate II. In the latter of these figures he is shown stretching his mouth, apparently yawning but actually preparing for an attack on another monkey behind the wire screen. Figure 7 of this plate indicates Skirrl in an interesting attitude of attention and with an obvious lack of self-consciousness. The same monkey is represented again in figures 8 and 9 of plate II, this time in the act of using hammer and saw.

All of the animals except the orang utan had been used more or less for experiments on behavior by Doctor Hamilton, but this prior work in no way interfered with my own investigation. Doctor Hamilton has accumulated a large mass of the most valuable and interesting observations on the behavior of monkeys, and he more thoroughly understands them than any other observer of whom I have knowledge. Much to my regret and embarrassment in connection with the present report, he has thus far published only a small portion of his data (Hamilton, 1911, 1914). In his most recent paper on "A study of sexual tendencies in monkeys and baboons," he has given important information concerning several of the monkeys which I have observed. For the convenience of readers who may make use of both his reports and mine, I am designating the animals by the names previously given them by Hamilton. The available and essential information concerning the individuals is presented below.

List of animals in collection Skirrl. Pithecus irus. Adult male.

Sobke. _P. rhesus_. Young adult male.

Gertie. _P. irus-rhesus_. Female. Born November, 1910.

Maud. _P. rhesus_. Young adult female.

Jimmy II. _P. irus_. Adult male.

Scotty. _P. irus_ (?). Adult male.

Tiny. _P. irus-rhesus_. Female. Born August, 1913.

Chatters. _P. irus_. Adult eunuch.

Daddy. _P. irus_. Adult eunuch.

Mutt. _P. irus_. Young adult male. Born August, 1911.

Julius. Pongo pygmaeus. Male. Age, 4 years to 5 years.

When I arrived in Santa Barbara, Doctor Hamilton was about to remodel, or rather reconstruct, his animal cages and laboratory. This gave us opportunity to adapt both to the special needs of my experiments. The laboratory was finally located and built in a grove of live oaks. From the front it is well shown by figure 10 of plate III, and from the rear, by figure 11. Its location was in every way satisfactory for my work, and in addition, the spot proved a delightful one in which to spend one's time.

Figure 12 is a ground plan, drawn to scale, of the laboratory and the adjoining cages, showing the relations of the several rooms of the laboratory among themselves and to the nine cages. Although the construction was throughout simple, everything was convenient and so planned as to expedite my experimental work. The large room A, adjoining the cages, was used exclusively for an experimental study of ideational behavior by means of my recently devised multiple-choice method. Additional, and supplementary, experiments were conducted in the large cage Z. Room D served as a store-room and work-shop.

The laboratory was forty feet long, twenty-two feet wide, and ten feet to the plate. Each small cage was six, by six, by twelve feet deep, while the large compartment into which each of the smaller cages opened was twenty-four feet long, ten feet wide, and twelve feet deep.

II

OBSERVATIONAL PROBLEMS AND METHODS

My chief observational task in Montecito was the study of ideational behavior, or of such adaptive behavior in monkeys and apes as corresponds to the ideational behavior of man. It was my plan to determine, so far as possible in the time at my disposal, the existence or absence of ideas and the role which they play in the solution of problems by monkeys and apes. I had in mind the behavioristic form of the perennial questions: Do these animals think, do they reason, and if so, what is the nature of these processes as indicated by the characteristics of their adaptive behavior?

My work, although obviously preliminary and incomplete, differs from most of the previous studies of the complex behavior of the infrahuman primates in that I relied chiefly upon a specially devised method and applied it systematically over a period of several months. The work was intensive and quantitative instead of more or less incidental, casual, and qualitative as has usually been the case. Naturally, during the course of my special study of ideational behavior observations were made relative to various other aspects of the life of my subjects. Such, for example, are my notes on the use of the hands, the instincts, the emotions, and the natural aptitudes of individuals. It is, indeed, impossible to observe any of the primates without noting most interesting and illuminating activities. And although the major portion of my time was spent in hard and monotonous work with my experimental apparatus, I found time each day to get into intimate touch with the free activities of my subjects and to observe their social relations and varied expressions of individuality. As a result of my close acquaintance with this band of primates, I feel more keenly than ever before the necessity of taking into account, in connection with all experimental analyses of behavior, the temperamental characteristics, experience, and affective peculiarities of individuals.

The light which I have obtained on the general problem of ideation has come, first, through a method which I have rather inaptly named the multiple-choice method, and second, and more incidentally, through a variety of supplementary methods which are described in Section IV of this report. These supplementary methods are simple tests of ideation rather than systematic modes of research. They differ from my chief method, among other respects, in that they have been used by various investigators during the past ten or fifteen years. It was not my aim to repeat precisely the observations made by others, but instead to verify some of them, and more especially, to throw additional light on my main problem and to further the analysis of complex behavior.

What has been referred to as the multiple-choice method was devised by me three years ago as a means of obtaining strictly comparable objective data concerning the problem-solving ability of various types and conditions of animals. The method was first tried with human subjects in the Psychopathic Hospital, Boston, with a crude keyboard apparatus which, however, proved wholly satisfactory as a means of demonstrating its value. It has since been applied by means of mechanisms especially adapted to the structure and activities of the organisms, to the study of the behavior of the crow, pig, rat, and ringdove (Yerkes, 1914; Coburn and Yerkes, 1915; Yerkes and Coburn, 1915). The method has also been applied with most gratifying results to the study of the characteristics of ideational behavior in human defectives,-- children, and adults,--and in subjects afflicted with various forms of mental disease. It is at present being tried out as a practical test in connection with vocational guidance and various forms of institutional examination, such as psychopathic hospital and court examinations.

As no adequate description of the method has yet been published to which I can here refer, it will be necessary to present its salient characteristics along with a description of the special form of apparatus which was found suitable for use with monkeys and apes.

The method is so planned as to enable the observer to present to any type or condition of organism which he wishes to study any one or all of a series of problems ranging from the extremely simple to the complex and difficultly soluble. All of the problems, however, are completely soluble by an organism of excellent ideational ability. For the human subject, the solution of the easiest problem of all requires almost no effort, whereas even moderately difficult problems may require many repetitions of effort and hours or days of application to the task. In each case, the solution of the problem depends upon the perception of a certain constant relation among a series of objects to which the subject is required to attend and respond. Such relations are, for example, secondness from one end of the group, middleness, simple alternation of ends, or progressive movement by constant steps from one end of a group to the other.

It is possible to present such relational problems by means of relatively simple reaction-mechanisms. In their essential features, all of the several types of multiple-choice apparatus designed by the writer and used either by him or by his students and assistants are the same. They consist of a series of precisely similar reaction-devices, any one or all of which may be used in connection with a given observation. These reaction-mechanisms are so chosen as to be suited to the structure and action-system of the animal to be studied. For the human being the mechanism consists of a simple key and the total apparatus is a bank of keys, with such electrical connections as are necessary to enable the observer to obtain satisfactory records of the subject's behavior. Let us suppose the bank of keys, as was actually the case in my first form of apparatus, to consist of twelve separate reaction-mechanisms; and let us suppose, further, the constant relation (problem) on the basis of which the subject is required to react to be that of middleness. It is evident that in successive trials or experiments the keys must be presented to the subject in odd groups, the possibilities being groups of 3, 5, 7, 9, or 11. If for a particular observation the experimenter wishes to present the first three keys at the left end of the keyboard, he pushes back the remaining nine keys so that they cannot be operated and requires the subject to select from the group of three keys the one which on being pressed causes a signal to appear. It is of course the clearly understood task of the subject to learn to select the correct key in the group on first trial. This becomes possible only as the subject observes the relation of the key which produces the desired effect to the other keys in the group. On the completion of a subject's reaction to the group of three keys, a group of seven keys at the opposite end of the keyboard may, for example, be presented. Similarly, the subject is required to discover with the minimum number of trials the correct reaction-mechanism. Thus, time after time, the experimenter presents a different group of keys so that the subject in no two successive trials is making use of the same portion of the keyboard. It is therefore impossible for him to react to spatial relations in the ordinary sense and manner, and unless he can perceive and appropriately respond to the particular relation which constitutes the only constant characteristic of the correct reaction-mechanism for a particular problem, he cannot solve the problem, or at least cannot solve it ideationally and on the basis of a small number of observations or trials.

For the various infrahuman animals whose ideational behavior has been studied by means of this method, it has been found eminently satisfactory to use as reaction-mechanisms a series of similar boxes, each with an entrance and an exit door. An incentive to the selection of the right box in a particular test is supplied by food, a small quantity of which is placed in a covered receptacle beyond the exit door of each of the boxes. Each time an animal enters a wrong box, it is punished for its mistake by being confined in that box for a certain period, ranging from five seconds to as much as two minutes with various individuals or types of organism. This discourages random, hasty, or careless choices. When the right box is selected, the exit door is immediately raised, thus uncovering the food, which serves as a reward. After eating the food thus provided, the animal, according to training, returns to the starting point and eagerly awaits an opportunity to attempt once more to find the reward which it has learned to expect. With this form of the apparatus, the boxes among which choice may be made are indicated by the raising (opening) of the front door.

Since with various birds and mammals the box form of apparatus had proved most satisfactory, I planned the primate apparatus along similar lines, aiming simply to adapt it to the somewhat different motor equipment and destructive tendencies of the monkeys. I shall now briefly describe this apparatus as it was constructed and used in the Montecito laboratory.

EXPLANATION OF PLATE IV

FIGURE 13.--Multiple-choice apparatus, showing observer's bench and writing stand. FIGURE 14.--Apparatus as seen from observer's bench. FIGURE 15.--Entrances to multiple-choice boxes as seen from the response-compartment. FIGURE 16.--Apparatus as seen from the rear, showing exit doors, food receptacles, and covers for same.

The apparatus was built in room A (figure 12), this room having been especially planned for it with respect to lighting as well as dimensions and approaches. It was unfortunately impossible to obtain photographs showing the whole of the apparatus, but it is hoped that the four partial views of plate IV may aid the reader who is unfamiliar with previously described similar devices to grasp readily the chief points of construction. In this plate, figure 13 shows the front of the complete apparatus, with the alleyway and door by way of which the experimenter could enter. The investigator's observation-bench and record-table also appear in this figure, together with weighted cords used to operate the various doors and the vertically placed levers by means of which each pair of doors could be locked. Figure 14 is the view presented to the observer as he stood on the bench or observation stand of figure 13 and looked over the entire apparatus. Three of the entrance doors are shown at the right of this figure as raised, whereas the remainder of the nine entrance doors of the apparatus are closed. Figure 15 is a view of the entrance doors from below the wire roof of the apparatus. Again, two of the doors are shown as raised, and three additional ones as closed. The rear of the apparatus appears in figure 16, in which some of the exit doors are closed and others open. In the latter case, the food receptacles appear, and on the lower part of the raised doors of the corresponding boxes may be seen metal covers for the food receptacles projecting at right angles to the doors, while on the lower edge of each door is an iron staple used to receive a sliding bar which could be operated from the observer's bench as a means of locking the doors after they had been closed. The space beyond the exit doors was used as an alleyway for the return of the animals to the starting point.

It will be necessary at various points in later descriptions to refer to these several figures. But further description of them will be more readily appreciated after a careful examination of the ground plan of the apparatus presented as figure 17 In accordance with the labelling of this figure, the experimenter enters the apparatus room through doorway 16, passes thence through doorways 17 and 10 to the large cage Z, from which he has direct access to the animals and can bring them into the apparatus. The multiple-choice mechanism proper, consisting of nine similar boxes (nine were used instead of twelve as a matter of convenience of construction, not because this smaller number is otherwise preferable) is labelled F. These boxes are numbered 1 to 9, beginning at the left. This numbering was adhered to in the recording of results throughout the investigation. The other important portions of the apparatus are the runway D, from which the subject at the experimenter's pleasure could be admitted through doorway 12 to the large response-chamber E; the alleyways G, H, and I, by way of which return to the starting point was possible; the observation bench C, with its approach step 13; and the observer's writing table A.

In the construction of this large apparatus, it was necessary to make provision for the extremely destructive tendencies of monkeys and anthropoid apes,--hence the apparent cumbersomeness of certain portions. It was equally necessary to provide for the protection of the observer and the prevention of escape of the subjects by completely covering the apparatus and alleyways with a heavy wire netting.

Each of the eighteen doors of the multiple-choice boxes, and in addition doors 11, 12, and 15 of the runway D, were operated by the observer from his bench C by means of weighted window cords which were carried by pulleys appropriately placed above the apparatus. Each weight was so chosen as to be just sufficient to hold its door in position after the experimenter had raised it. For the convenience of the experimenter in the rapid operation of the twenty-one doors, the weights for the doors of runway D were painted gray, those for the entrance doors, white, and those for the exit doors, black.

In each entrance door, as is shown in figure 15 of plate IV, a window was cut so that the experimenter might watch the animal after it had entered a given box, and especially note when it left the box after having received its reward. This window was covered with wire netting. No such windows were necessary in the exit doors, but to them were attached heavy galvanized iron flanges which served to cover the food receptacles. One of these flanges is labelled o in figure 17. The food receptacles were provided by boring holes in a 2 by 4 inch timber securely nailed to the floor immediately outside of the exit doors. Into these holes aluminum cups fitted snugly, and the iron flanges, when the doors were closed, fitted so closely over the cups that it was impossible for the animals to obtain food from them.

A, record stand; C, bench for observer; B, step as approach to C; D, alleyway leading to E, response-compartment; F, one of the nine (1-9) similar multiple-choice boxes; G, H, alleyways leading from boxes to starting point at D; I, alleyway used by experimenter as approach to rear of apparatus; W, W, windows; P, alleyway; Z, large cage; 16, entrance to room A; 17, entrance to apparatus and thence via 10 to cages; 18, entrance to alleyway 1; 11, 15, entrances to D; 12, entrance to E; 13, entrance door of box 5; 14, exit door of box 5; o, cover for food receptacle.]

As originally constructed, no provision was made in the apparatus for locking the entrance and exit doors of the several boxes when they were closed. But as two of the subjects after a time learned to open the doors from either outside or inside the boxes, it became necessary to introduce locking devices which could be operated by the experimenter from the observation bench. This was readily accomplished by cutting holes in the floor, which permitted an iron staple, screwed to the lower edge of each door, to project through the floor. Through these staples by means of a lever for each of the nine boxes, the observer was able to slide a wooden bar, placed beneath the floor of the room, thus locking or unlocking either the entrance door, the exit door, or both, in the case of any one of the nine boxes.

Since figure 17 is drawn to scale, it will be needless to give more than a few of the dimensions of the apparatus. Each of the boxes was 42 inches long, 18 inches wide, and 72 inches deep, inside measurements. The alleys D, I, and H were 24 inches, and G 30 inches wide, by 6 feet deep. The doors of the several boxes were 18 inches wide, by 5 feet high, while those in the alleyways were 24 inches wide by 6 feet high. The response-compartment E of figure 17 was 14 feet 4 inches, by 8 feet, by 6 feet in depth. In order that the apparatus might be used with adult human subjects conveniently, if such use should prove desirable, the depth throughout was made 6 feet, and it was therefore possible for the experimenter to walk about erect in it.

The experimental procedure was briefly as follows: A small quantity of food having been placed in each of the food cups and covered by the metal flanges on the exit doors, the experimenter raised door 11 of figure 17 and then opened door 10 and the door of the cage in which the desired subject was confined. After the latter, in search of food, had entered the runway D, the experimenter lowered door 11 to keep it in this runway, and immediately proceeded to set the reaction-mechanisms for an experiment (trial). Let us suppose that the first setting to be tried involved all of the nine boxes. Each of the entrance doors would therefore be raised. Let us further suppose that the right door is defined as the middle one of the group. With the apparatus properly set, the experimenter next raises door 12, thus admitting the animal to the response-compartment E. Any one of the nine boxes may now be entered by it. But if any except number 5, the middle member of the group, be entered, the entrance door is immediately lowered and both the exit and entrance doors locked in position so that the animal is forced to remain in the box for a stated period, say thirty seconds. At the expiration of this time the entrance door is raised and the animal allowed to retrace its steps and make another choice. When the middle box is chosen, the entrance door is lowered and the exit door immediately raised, thus uncovering the food, which the animal eats. As a rule, by my monkeys and ape the reward was eaten in the alleyway G instead of in the multiple-choice box. As soon as the food has been eaten, the exit door is lowered by the experimenter, and the animal returns by way of G and H to runway D, where it awaits its next trial.

As rewards, bananas and peanuts were found very satisfactory, and although occasionally other foods were supplied in small quantities, they were on the whole less constantly desired than the former.

Four problems which had previously been presented to other organisms were in precisely the same form presented to the three primates. These problems may be described, briefly, by definition of the right reaction mechanism, thus: problem 1, the first mechanism at the subject's left; problem 2, the second mechanism at the subject's right (that is, from the end of the series at the subject's right); problem 3, alternately, the first mechanism at the subject's left and the first at its right; problem 4, the middle mechanism of the group.

It was my intention to present these four problems, in order, to each of the three animals, proceeding with them as rapidly as they were solved. But as it happened, only one of the three subjects got as far as the fourth problem. When observations had to be discontinued, Sobke was well along with the last, or fourth problem; Skirrl was at work at the third problem; and Julius had failed to solve the second problem.

For each of the problems, a series of ten different settings of the doors was determined upon in advance. These settings differ from those employed in a similar investigation with the pig only in that the numbering of the doors is reversed. In the present apparatus, the boxes as viewed from the front (entrance) are numbered from the left to the right end, whereas those of the pig apparatus were numbered from the right end to the left end.

Below are presented for each of the several problems (1) the numbers of the settings presented in series; (2) the numbers of the doors open; (3) the number of doors open in each setting and for the series of ten settings; and (4) the number of the right door.

PROBLEM 1. First mechanism at left of group

Doors	No. of doors open	No. of Settings	open doors	open right door
1	1.2.3	3	1	
2	8.9	2	8	
3	3.4.5.6.7	5	3	
4	7.8.9	3	7	
5	2.3.4.5.6	5	2	
6	6.7.8	3	6	
7	5.6.7	3	5	
8	4.5.6.7.8	5	4	
9	7.8.9	3	7	
10	1.2.3	3	1 -- Total 35	

PROBLEM 2. Second mechanism from the right end of group

Doors	No. of doors open	No. of Settings	open doors	open right door
1	7.8.9	3	8	
2	1.2.3.4	4	3	
3	2.3.4.5.6.7	6	6	

Doors	No. of Settings	No. of open doors	open right door
4	1.2.3.4.5.6	6	5
5	4.5.6.7.8	5	7
6	1.2.3	3	2
7	2.3.4.5	4	4
8	1.2.3.4.5.6.7.8.9	9	8
9	1.2.3.4	4	3
10	3.4.5.6.7.8	6	7 -- Total 50

PROBLEM 3. Alternately the first mechanism at the left and the first at the right end of the group

Doors	No. of Settings	No. of open doors	open right door
1	5.6.7	3	5
2	5.6.7	3	7
3	1.2.3.4.5.6	6	1
4	1.2.3.4.5.6	6	6
5	4.5.6.7.8	5	4
6	4.5.6.7.8	5	8
7	2.3.4.5	4	2
8	2.3.4.5	4	5
9	3.4.5.6.7.8.9	7	3
10	3.4.5.6.7.8.9	7	9 -- Total 50

PROBLEM 4. Middle mechanism of the group

Doors	No. of Settings	No. of open doors	open right door
1	2.3.4	3	3
2	5.6.7.8.9	5	7
3	1.2.3.4.5.6.7	7	4
4	7.8.9	3	8
5	4.5.6.7.8	5	6
6	1.2.3.4.5.6.7.8.9	9	5
7	1.2.3	3	2
8	2.3.4.5.6	5	4
9	3.4.5.6.7.8.9	7	6
10	6.7.8	3	7 -- Total 50

It was found desirable after a problem had been solved to present a new and radically different series of settings in order to determine to what extent the subject had learned to choose the correct door by memorizing each particular setting. These supplementary observations may be known as control experiments, and the settings as supplementary settings. In case of these, as for the original settings, the essential facts are presented in tabular arrangement.

Settings for Control Experiments

PROBLEM 1. First at left end

Doors	No. of Settings	No. of open doors	open right door
1	2.3.4	3	2
2	6.7.8.9	4	6
3	3.4.5	3	3
4	4.5.6.7.8.9	6	4
5	6.7.8.9	4	6
6	1.2.3.4.5	5	1
7	2.3.4.5.6.7.8	7	2
8	3.4.5.6.7.8	6	3
9	5.6.7	3	5
10	1.2.3.4.5.6.7.8.9	9	1

PROBLEM 2. Second from right end

Doors	No. of Settings	No. of open doors	open right door
1	5.6.7.8	4	7
2	2.3.4.5.6	5	5
3	1.2.3.4.5.6.7.8.9	9	8
4	5.6.7	3	6
5	1.2.3.4	4	3
6	4.5.6	3	5
7	2.3.4.5	4	4
8	1.2.3	3	2
9	1.2.3.4.5.6.7	7	6
10	2.3.4.5.6.7.8.9	8	8

PROBLEM 3. Alternate left and right ends

Doors	No. of	No. of Settings open doors	open right door
1	5.6	2	5
2	5.6	2	6
3	4.5.6.7.8.9	6	4
4	4.5.6.7.8.9	6	9
5	1.2.3.4.5	5	1
6	1.2.3.4.5	5	5
7	2.3.4.5.6.7	6	2
8	2.3.4.5.6.7	6	7
9	3.4.5.6.7.8	6	3
10	3.4.5.6.7.8	6	8

PROBLEM 4. Middle

Doors	No. of	No. of Settings open doors	open right door
1	4.5.6.7.8	5	6
2	1.2.3	3	2
3	1.2.3.4.5.6.7.8.9	9	5
4	2.3.4.5.6	5	4
5	6.7.8	3	7
6	3.4.5.6.7.8.9	7	6
7	7.8.9	3	8
8	1.2.3.4.5.6.7	7	4
9	2.3.4	3	3
10	3.4.5.6.7	5	5

It was my aim so far as possible to present to a given subject each day the ten settings under a given problem in order, without interruption. If for any reason the series of observations had to be interrupted, it was resumed at the same point subsequently. Occasionally it was found desirable or necessary to present only five of the series of ten settings in succession and then to interrupt observations for an interval of a few minutes or even several hours. But as a rule it was possible to present the series of ten settings. All things being considered, it proved more satisfactory to give only ten trials a day to each subject. Frequently twenty and rarely thirty trials were given on the same day. In such cases the series of settings was simply repeated. The only pause between trials was that necessary for resetting the entrance doors and replenishing the food which served as a reward for success.

III

RESULTS OF MULTIPLE-CHOICE EXPERIMENTS

1. Skirrl, Pithecus irus _Problem 1. First at the Left End_

Systematic work with the multiple-choice apparatus and method described in the previous section was undertaken early in April with Skirrl, Sobke, and Julius. The results for each of them are now to be presented with such measure of detail as their importance seems to justify.

Skirrl had previously been used by Doctor Hamilton in an experimental study of reactive tendencies. He proved so remarkably inefficient in the work that Doctor Hamilton was led to characterize him as feeble-minded, and to recommend him to me for further study because of his mental peculiarities. With me he was from the first frank, aggressive, and inclined to be savage. It was soon possible for me to go into the large cage, Z, with him and allow him to take food from my hand. He was without fear of the experimental apparatus and it proved relatively easy to accustom him to the routine of the experiment. Throughout the work he was rather slow, inattentive, and erratic.

Beginning on April 7, I sought to acquaint him with the multiple-choice apparatus by allowing him to make trips through the several boxes, with the reward of food each time. Thus, for example, with the entrance and exit doors of box 7 raised, the monkey was allowed to pass into the reaction-compartment E and thence through box 7 to the food cup. As soon as he had finished eating, he was called back to D by the experimenter and, after a few seconds, allowed, similarly, to make a trip by way of one of the other boxes. By reason of this preliminary training he soon came to seek eagerly for the reward of food.

On April 10 the apparatus was painted white in order to increase the lightness and thus render it easier for the experimenter to observe the animal's movements, and when on April 12 Skirrl was again introduced to it for further preliminary training, he utterly refused to enter the boxes, giving every indication of extreme fear of the white floors and even of the sides of the boxes. Finally, the attempts to induce him to enter the boxes had to be given up, and he was returned to his cage unfed. The following day I was equally unsuccessful in either driving or tempting him with food into the apparatus. But on April 14 he was so hungry that he was finally lured in by the use of food. He cautiously approached the boxes and attempted to climb through on the sides instead of walking on the floor. It was perfectly evident that he had an instinctive or an acquired fear of the white surfaces. As the matter was of prime importance for the success of my work, I inquired of Doctor Hamilton, and of the men in charge of the cages, for any incident which might account for this peculiar behavior, and I learned that some three months earlier, while the animal cages were being whitewashed, Skirrl had jumped at one of the laborers who was applying a brush to the framework of one of the cages and had shaken some lime into his eyes. He was greatly frightened and enraged. Evidently he experienced extreme discomfort, if not acute pain, and there resulted an association with whiteness which was quite sufficient to cause him to avoid the freshly painted apparatus.

Having obtained an adequate explanation of this monkey's peculiar behavior, I proceeded with my efforts to induce him to work smoothly and rapidly, and on April 15, by covering the floor with sawdust, I so diminished the influence of the whiteness as to render the preliminary training fairly satisfactory. At the end of two more days everything was going so well that it seemed desirable to begin the regular experiment.

On the morning of April 19, Skirrl was introduced to the apparatus and given his first series of ten trials on problem 1. This problem demanded the selection of the first door at the left in any group of open doors. The procedure was as previously described in that the experimenter raised the entrance doors of a certain group of boxes, admitted the animal to the reaction-chamber, punished incorrect choices by confining the animal for thirty seconds, and rewarded correct choices by raising the exit door and thus permitting escape and the obtaining of food. The trials were given in rapid succession, and the total time required for this first series of ten trials was thirty-five minutes. Skirrl worked faithfully throughout this interval and exhibited no marked discouragement. When confined in a box he showed uneasiness and dissatisfaction by moving about constantly, shaking the doors, and trying to raise them in order to escape.

For the series of settings used in connection with problem 1, the reader is referred to page 18. In the first setting, the doors numbered 1, 2, and 3, were opened. As it happened, the animal when admitted to the reaction-chamber immediately chose box I. Having received the reward of food, he was called back to D, and doors 8 and 9 having been raised in preparation for the next trial, he was again admitted to the reaction-chamber. This time he quickly chose box 9 and was confined therein for thirty seconds. On being released, he chose after an interval of four minutes, box 8, thus completing the trial.

As it is highly important, not only in connection with the present description of behavior, but also for subsequent comparison of the reactions of different types of organism in this experiment, to present the detailed records for each trial, tables have been constructed which offer in brief space the essential data for every trial in connection with a given problem.

Table 1 contains the results for Skirrl in problem 1. It is constructed as follows: the date of a series of trials appears in the first vertical column; the numbers (and number) of the trials for the series or date appear in column 2; the following ten columns present respectively the results of the trials for each of the ten settings. Each number, in these results, designates a box entered. At the extreme right of the table are three columns which indicate, first, the number of trials in which the right box was chosen first, column headed R; and second, the number of trials in which at least one incorrect choice occurred, column headed W. In the last column, the daily ratio of these first choices appears.

Taking the first line of table 1 below the explanatory headings, we note on April 19 ten trials, numbered 1 to 10, were given to Skirrl. In trial 1, with setting 1, he chose correctly the first time, and the record is therefore simply 1. In trial 2, setting 2, he incorrectly chose box 9, the first time. At his next opportunity, he chose box 8, which was the right one. The record therefore reads 9.8. In trial 3, setting 3, he chose incorrectly twice before finally selecting the right box. The record reads 6.7.3, and so on throughout the ten trials which constitute a series. The summary for this series indicates three right and seven wrong first choices, that is, three cases in which the right box was entered first. The ratio of right to wrong first choices is therefore 1 to 2.33. Since the total number of doors open in the ten settings is thirty-five, and since in each of the ten settings one door is describable as the right door, the probable ratio, apart from the effects of training, of right to wrong first choices is 1 to 2.50. It is evident, therefore, that Skirrl in his first series of trials closely approximated expectation in the number of mistakes.

TABLE 1

Results for Skirrl, _P. irus_, in Problem 1

========+==========+=========+=========+===========+===========
==+===========+===========+===========+==============+===========
=+===========+===+===+======== | No. | S.1 | S.2 | S.3 | S.4 | S.5 | S.6 | S.7 | S.8 | S.9 | S.10 | | | Ratio Date | of | | | | | | | | | | | R | W | of | trials | 1.2.3 | 8.9 | 3.4.5.6.7 | 7.8.9 | 2.3.4.5.6 | 6.7.8 | 5.6.7 | 4.5.6.7.8 | 7.8.9 | 1.2.3 | | | R to W --------+----------+---------+---------+-----------+--------------+---------
--+-----------+-----------+-----------+--------------+-----------+-----------+---+---+-------- April | | | |
| | | | | | | | | 19 | 1- 10 | 1 | 9.8 | 6.7.3 | 9.7 | 6.2 | 7.8.6 | {6.7.7.7 | 4 | 7
| 2.3.3.1 | 3 | 7 | 1:2.33 | | | | | | | | {6.5 | | | | | | | | | | | | | | | | | | | | | | | | | | | |
20 | 11- 20 | 3.2.1 | 9.8 | 5.3 | 7 | 4.2 | 8.8.6 | 5 | 8.4 | 7 | 3.1 | 3 | 7 |
1:2.33 21 | 21- 30 | 3.1 | 8 | 3 | 8.7 | 6.2 | 6 | 5 | 6.4 | 9.7 | 1 | 5 | 5 | 1:1.00
22 | 31- 40 | 1 | 9.8 | 3 | 7 | 6.2 | 6 | 6.7.5 | 5.8.4 | 9.8.9.8.7 | 2.1 | 4 | 6 |
1:1.50 23 | 41- 50 | 2.3.1 | 8 | 5.7.3 | 7 | 4.2 | 6 | 5 | 7.8.4 | 7 | 3.1 | 5 | 5 |
1:1.00 24 | 51- 60 | 1 | 8 | 4.5.7.3 | 9.7 | 5.6.2 | 6 | 6.7.5 | 6.4 | 8.9.7 | 1 | 4
| 6 | 1:1.50 26 | 61- 70 | 1 | 8 | 6.7.4.7.3 | 7 | 4.5.6.2 | 6 | 5 | 8.4 | 7 |
3.2.3.1 | 6 | 4 | 1: .67 27 | 71- 80 | 3.1 | 8 | 3 | 9.7 | 4.6.2 | 7.6 | 6.5 | 5.8.4 |
7 | 1 | 4 | 6 | 1:1.50 28 | 81- 90 | 2.3.1 | 8 | 3 | 7 | 4.5.6.2 | 6 | 5 | 5.8.4 | 7
| 1 | 7 | 3 | 1: .43 29 | 91- 100 | 1 | 8 | 3 | 9.7 | 6.2 | 6 | 5 | 4 | 7 | 1 | 8 | 2 |
1: .25 30 | 101- 110 | 1 | 8 | 4.3 | 7 | 5.6.2 | 6 | 5 | 4 | 7 | 2.3.1 | 7 | 3 | 1:
.43 May | | | | | | | | | | | | | | 1 | 111- 120 | 2.3.2.1 | 8 | 3 | 7 | 2 | 6 | 5 |
4 | 7 | 1 | 9 | 1 | 1: .11 3 | 121- 130 | 1 | 8 | 5.6.3 | 7 | 4.5.2 | 6 | 5 | 4 | 7 |
1 | 8 | 2 | 1: .25 4 and 5 | 131- 140 | 3.2.1 | 8[1] | 3 | 7 | 2 | 6 | 5 | 4 | 7 | 1
| 9 | 1 | 1: .11 5 | 141- 150 | 1 | 8 | 4.3 | 7 | 2 | 6 | 5 | 4 | 7 | 1 | 9 | 1 | 1:
.11 --------+----------+---------+---------+-----------+--------------+-----------+-----------+-----------+---
--------+--------------+-----------+-----------+-----------+---+---+-------- | | | | | | | | 2.3.4.5 | | |
1.2.3.4.5 | | | | | | 2.3.4 | 6.7.8.9 | 3.4.5 | 4.5.6.7.8.9 | 6.7.8.9 | 1.2.3.4.5 |
6.7.8 | 3.4.5.6.7.8 | 5.6.7 | 6.7.8.9 | | | | +---------+----------+-----------+-------------
-+-----------+-----------+-----------+--------------+-----------+-----------+---+---+-------- 6 |
1- 10 | 2 | 6 | 3 | 4 | 6 | 3.2.1 | 6.2 | 5.6.7.8.3 | 5 | 6.1 | 6 | 4 | 1: .67

```
========+==========+=========+=========+==========+==========
==+==========+==========+==========+============+=========
=+==========+===+===+========
```

[Footnote 1: End of series on May 4.]

By reading downward in any particular column of results, one obtains a description of the changes in the animal's reaction to a particular setting of the doors. Thus, for instance, in the case of setting 1, which was presented to the animal in trials numbered 1, 11, 21, and so on to 141, it is clear from the records that no definite improvement occurred. But oddly enough, in the case of setting 10, which presented the same group of open doors, almost all of the reactions are right in the lower half of the column. For setting 2, it is evident that mistakes soon disappeared.

Comparison of the data of table 1 indicates that the number of correct first choices is inversely proportional to the number of doors in use, while the number of choices made in a given trial is directly proportional to the number of doors in use.

During the first week of work on this problem, Skirrl improved markedly. His performance was somewhat irregular and unpredictable, but on the whole the experiment seemed fairly satisfactory. Cold, cloudy, or rainy days tended to diminish steadiness and to increase the number of mistakes. Similarly, absence of hunger was unfavorable to continuous effort to find the right box.

The period of confinement, as punishment for wrong choices, was increased from thirty seconds to sixty seconds on April 26. But there is no satisfactory evidence that this favored the solution of the problem. Work on May 4 was interrupted by a severe storm, the noise of which so distracted the monkey that he ceased to work. Consequently, observations were interrupted on the completion of trial 132, and on May 5, the series was begun with setting 3. On this date, eighteen trials were given in succession, and in only one of them did a mistake occur. Since the ten trials numbered 133 to 142 were correct, Skirrl was considered to have solved problem 1, and systematic training was discontinued.

On the following day, as a measure of the extent to which the animal had learned to select the first door at the left no matter what its position or the number of doors in the group presented, a control series was given in which the settings differed from the regular series of settings. These supplementary settings are presented at the bottom of table 1 together with the records of reaction in ten trials.

Since in only six of these ten control settings was the first choice correct, it is scarcely fair to insist that the animal was reacting on the basis of an ideational solution of the problem. Rather, it would seem that he had learned to react to particular settings. A careful study of all of the data of response, together with notes on the varied behavior of the animal during the experiments, justifies the statement that Skirrl's solution of problem 1 was incomplete and unreliable. It was highly dependent upon the particular situation, or even the particular door at the left end of the group, and slightly if at all dependent upon anything comparable to the human idea of first at the left of the group.

This particular series of observations has been described and discussed in some detail in order to make the chief points of method clear. It will be needless, hereafter, to refer explicitly to many of the characteristics of reaction or to the important points in the construction of tables which have been mentioned.

A graphic representation of Skirrl's learning process in problem 1 is presented in figure 18. The irregularities are most striking, and fairly indicate the erraticness of the animal. The curve is based upon the data in next to the last column of table 1, that is, the column presenting the errors or wrong first choices in each series of trials.

Unquestionably, the form of such a curve of learning should be considered in connection with the method or methods of selecting the right box employed by the animal during the course of experimentation. It appears from an analysis of the behavior of Skirrl in problem 1 that there developed a single definite and persistent method, namely, that of going to one box in the group, and in case it happened to be a wrong one, of choosing, on emergence from it, the next toward the right end of the group, and so on down the line. Having reached the extreme right end, the tendency was to follow the side of the reaction-chamber around to the opposite end and to enter the first box at the left end of the group, which was, of course, the right one. This method appears, with certain slight variations, in approximately ninety per cent of the trials which involved incorrect choices. Thus, in the case of trials 121 to 130, of which eight exhibit right first choices, the remaining two exhibit the method described above except that the final member at the right end of the group was in each case omitted.

On the whole, Skirrl's behavior in connection with this problem appears to indicate a low order of intelligence. He persisted in such stupid acts as that of turning, after emergence from the right box, toward the right and passing into the blind alley I, instead of toward the left, through G and H, to D. In contrast with the other animals, he spent much time before the closed doors of the boxes, instead of going directly to the open doors, some one of which marked the box in which the reward of food could be obtained. It is, moreover, obvious that his responses, as they appear in table 1, are extremely different from those of a human being who is capable of bringing the idea of first at the left end to bear upon the problem in question.

Problem 2. Second from the Right End

Following the series of control trials of problem 1 given to Skirrl on May 6, a period of four days was allowed during which the animal was merely fed in the boxes each day. This was done in order that he should partially lose the effects of his previous training to choose the first box at the left before being presented with the second problem, the second box from the right.

On May 11 regular experimentation was begun with problem 2. Naturally the situation presented unusual difficulties to the monkey because of his previously acquired habit, and on the first day it was possible to give only five trials, in all except the first of which Skirrl had to be aided by the experimenter to find the right box. He persistently, as appears in the first line of records of table 2, entered the first box at the left. The series was continued on May 13, but with very unsatisfactory results, since he apparently had been greatly discouraged by the unusual difficulties previously met. Only four trials could be given, and in these the showing made was very poor. It is noteworthy, however, that in trials 6, 7, and 8, May 13, there was no marked tendency to choose the first box at the left. Thus quickly had the force of the previous habit been broken.

For problem 2, the total number of open doors in the ten settings is fifty, as appears from the data on page 18, and as ten of these fifty open doors may be defined as right ones, the expected ratio of right to wrong first choices in the absence of previous training is 1 to 4. The actual ratio for the first series given in problem 2 is 1 to 8, while in the second series it is 0 to 10.

On the morning of May 13, work was interrupted in the ninth trial by what seemed at the moment a peculiarly unfortunate accident, but in the light of later developments, an incident most fruitful of valuable results.

Skirrl, in trial 9, directly entered box 1. Since this was not the right box, he was punished by being confined in it for ten seconds. While in the box he howled and when the entrance door was raised for him to retrace his steps, he came out with a rush, showing extreme excitement and either rage or fear, I could not be sure which. At intervals he uttered loud cries, which I am now able to identify as cries of alarm. Repeatedly he went to the open door of box 1 and peered in, or peered down through the hole in the floor which received the staple on the door. He refused to enter any one of the open boxes and continued, at intervals of every half minute or so, his cries. For thirty minutes I waited, hoping to be able to induce him to complete the series of trials, but in vain. Although it was obvious that he was eager to escape from the apparatus, he would not enter any of the boxes even when the exit doors were raised. Instead, he gnawed at the door (12 in fig. 17) to the alleyway D and attempted to force his way through, instead of taking the easy and clear route to the alleys, through one of the boxes. His behavior was most surprising and puzzling. Finally, I gave up the attempt to complete the series and returned him to his cage by way of the entrance door to the response-compartment E.

I then entered the apparatus to seek some explanation of the animal's behavior, and my search was rewarded by the finding of two sharp pointed nails which protruded for an inch or more in the middle of the floor of box 1. My assistant, who had been charged with the task of installing the locks for the several doors, had used nails instead of screws for attaching staples underneath the floor and had neglected to clinch the nails. Skirrl, in the dim light of the box, doubtless stepped upon one of the nails and inflicted a painful, although not serious, injury upon himself. It was impossible for him to see clearly the source of his injury. He was greatly frightened and expressed the emotion most vigorously. His behavior strongly suggested a superstitious dread of some unseen danger. It may be that the instinctive fear of snakes, so strong in monkeys, was partly responsible for his response.

The first result of this accident was that more than two weeks were lost, for it was impossible, during the next few days, to induce the animal to enter any of the multiple-choice boxes voluntarily. From May 14 to May 24, I labored daily to overcome his newly acquired fear. The usual procedure was to coax him through one box after another by standing at the exit door with some tempting morsel of food. After several days of this treatment, he again trusted himself to the boxes, although very circumspectly and only when both entrance and exit doors were raised. Not until May 24 was it possible to resume regular experimentation, and on that day it was found necessary to indicate the right box by raising the exit door slightly and then immediately lowering it. Trials in which this form of aid was given are indicated in table 2 by a star following the last choice.

Gradually, Skirrl regained his confidence in the apparatus and began to work more naturally. For a long time he would not stand punishment, and it was necessary for the experimenter to be very careful in locking the doors, since the sound of the bar sliding beneath the floor often frightened and caused him to quit work. Day after day the tendency to peer through the holes in the floor at the entrance to the boxes rendered it clear that the animal feared some danger from beneath the floor. This behavior was so persistent that much time was wasted in the experiments.

On the last day of May, punishment by confinement for ten seconds in wrong boxes was introduced, but since this tended to discourage the monkey, there was substituted for it on June 1 the punishment of forcing him to work his way out of each wrong box by raising the entrance door which had been closed behind him. This he could fairly readily do, and his stay in a box rarely measured more than ten seconds.

As a variation in the mode of procedure, confinement for thirty seconds was tried on June 5, but it worked unsatisfactorily and had to be abandoned. During this series, the animal was startled by the sound from one of the sliding bars under the floor, and in the sixth trial he refused to work.

As improvement was very slow, varied modes of rewarding and punishing the animal were tried in the hope of discovering a means of facilitating the work. Among the former are the use of banana, grapes, peanuts, and other eagerly sought foods in varying quantities, and in the latter are included periods of confinement ranging from ten seconds to sixty seconds. In the end, confinement of about thirty seconds, combined with a small quantity of food which was much to the monkey's taste, gave most favorable results.

All this time Skirrl's attention to the task in hand was seldom good. He was easily diverted and even when extremely hungry, often stopped work in the middle of an early trial, yawned repeatedly and finally sat down to wait for release from the apparatus.

The results obtained during the long continued trials with this animal in problem 2 are presented in table 2, which differs from the previously described table, first, in that several of the trials are followed by an asterisk to indicate that aid was given by the experimenter, and second, in that two additional columns, headed, respectively, R and W, are presented. These give the right and wrong first choices for each day, whereas the two columns preceding them give the same data for each series of ten trials. Similarly, the ratio of right to wrong choices is presented for each day in table 2, instead of for each series of ten trials as in table 1.

From the results of table 2, several peculiarly interesting facts appear. In the first place the influence of the habit of choosing the first box at the left disappears with surprising suddenness, and in the second place, there are remarkable contrasts in the results for different settings as they appear in their respective vertical columns. Thus, in the case of setting 1, after the first trial mistakes became relatively infrequent, whereas in setting 6, which involved the same number of doors, mistakes continued to be the rule until nearly a thousand trials had been given. The most likely explanation of this difference is that for some reason the animal avoided box 9.

The _reactive tendencies_, or better, the methods of reaction which manifested themselves during this long series of observations may be described as follows: (a) choice of the first box at the left; (b) random choice with tendency to choose first, a box near the middle of the group; (c) choice of first box at the right followed by the one next to it on the left; (d) direct choice of the right box.

TABLE 2

Results for Skirrl, _P. irus_, in Problem 2

```
========+===========+=============+=============+============
==+=============+=============+=============+=============+
=============+=============+=============+===+===+===+===+==
====== | No. | S.1 | S.2 | S.3 | S.4 | S.5 | S.6 | S.7 | S.8 | S.9 | S.10 | | | |
Ratio Date | of | | | | | | | | 1.2.3.4.5 | | | R | W | R | W | of | trials | 7.8.9 |
1.2.3.4 | 2.3.4.5.6.7 | 1.2.3.4.5.6 | 4.5.6.7.8 | 1.2.3 | 2.3.4.5 | 6.7.8.9 |
1.2.3.4 | 3.4.5.6.7.8 | | | | | R to W --------+-----------+-------------+-------------+--
------------+-------------+-------------+-------------+-------------+-------------+-------------
---+-------------+---+---+---+---+-------- May | | | | | | | | | | | | | | | | | 11&13 |
1- 9 | 7.7.9.7.8 | {1.2.2.1.4.1 | {2.3.2.3.2.5 | {4.6.1.4.1.1 | 4.4.7 | 3.1.2 | 4 |
4.1.8 | 1 | | 1 | 8 | 1 | 8 | 1: 8.00 | | | {2.1.2.1.3 | {2.3.2.5.6 | {2.6.1.6.5 | | |
| | | | | | | | | | | | | | | | | | | | | | | 24 | 11- 20 | 8*[1] | 2.4.3* | 4.5.6* |
2.2.5* | 5.6.6.7* | 3.1.2 | {5.2.3.5.3.2 | 4.6.8* | 4.4.3* | 5.5.6.7* | 0 |10 | 0
|10 | 0:10.00 | | | | | | | | {3.5.2.4* | | | | | | | | | | | | | | | | | | | | | | |
25 | 21- 30 | 8* | 4.4.3* | 5.6 | {6.6.2.3.4 | 6.7 | 2 | 4 | 5.6.3.8 | 4.4.3 |
6.4.6.8.7 | 2 | 8 | 2 | 8 | 1: 4.00 | | | | | {6.6.5* | | | | | | | | | | | | | | |
| | | | | | | | | | 26 | 31- 40 | 8 | 4.3 | 6 | 4.5 | 6.7 | 3.2 | 5.4 | 5.8 | 4.3 |
5.3.8.7 | 2 | 8 | 2 | 8 | 1: 4.00 | | | | | | | | | | | | | | | | | 27 | 41- 50 | 8 |
4.4.3 | 6 | 5 | 6.8.6.8.7 | 3.3.3.2 | 5.4 | {6.5.4.3 | 4.3 | 5.4.8.7 | 3 | 7 | 3 | 7 |
1: 2.33 | | | | | | | | | | {2.1.5.8 | | | | | | | | | | | | | | | | | | | | | 28 | 51-
60 | 8 | 4.4.3 | 7.6 | 5 | 5.6.7 | 3.3.3.2 | 4 | {5.4.3 | 4.3 | {5.4.3.3.4.5 | 3 | 7 |
3 | 7 | 1: 2.33 | | | | | | | | | {3.6.8 | | {6.4.3.5.7 | | | | | | | | | | | | | |
| | | | 29 | 61- 70 | 8 | 4.3 | 6 | 6.6.5 | 7 | 3.3.3.2 | 5.4 | 7.6.4.7.6.8 | 4.3 | 7
| 4 | 6 | 4 | 6 | 1: 1.50 | | | | | | | | | | | | | | | | | | 31 | 71- 80 | 8 | 4.4.4.3 | 6
| 6.5 | 6.8.7 | 3.2 | 5.4 | {6.7.6.4.3 | 4.3 | 6.7 | 2 | 8 | 2 | 8 | 1: 4.00 June | |
| | | | | | | {2.6.3.7.8 | | | | | | | | | | | | {6.8.6.5.4 | | | | | | | | | | 1 | 81-
90 | 8 | 4.3 | 6 | 5 | {6.5.6.5.8 | 3.1.3.2 | 5.4 | 8 | 4.3 | 7 | 5 | 5 | | | | | | | |
| {5.4.6.4.7 | | | | | | | | | | | | | | | | | | | | | | | | | | | | | | " | 91- 100 | 9.7.8 |
4.2.4.3 | 7.5.6 | 5 | 6.8.7 | 3.3.1.2 | 5.3.4 | 8 | 4.3 | 6.8.7 | 2 | 8 | 7 |13 | 1:
1.86 2 | 101- 110 | 8 | 4.3 | 6 | 5 | 7 | 3.2 | 5.4 | 7.8 | 4.3 | 6.8.6.5.7 | 4 | 6 |
| | | | | | | | | | | | | | | | | | | " | 111- 120 | 8 | 4.3 | 7.3.5.7.6 | {6.2.3.6.4 | 7
```

| 3.2 | {5.2.3.5.3.2 | 9.6.4.7.8 | {4.1.2 | 6.8.7 | 2 | 8 | 6 |14 | 1: 2.33 | | | | |
{3.6.2.5 | | | {3.5.2.3.4 | | {4.2.3 | | | | | | | | | | | | | | | | | {6.8.6.3 | | | | |
3 | 121- 130 | 8 | 4.4.3 | 6 | 5 | 6.7 | 3.2 | {5.3.2.3 | 8 | 4.2.3 | {5.4.5.8.8 | 4
| 6 | | | | | | | | | | | {5.2.5.4 | | | {6.3.8.7 | | | | | | | | | | | | | | | | | | | |
| " | 131- 140 | 8 | 4.3 | 5.7.3.2.6 | 4.5 | 5.7 | 1.3.2 | 5.3.4 | 6.7.8 | 4.2.1.3 |
7 | 2 | 8 | 6 |14 | 1: 2.33 | | | | | | | | | | | | | | | | | 4 | 141- 150 | 8 | 4.3 |
7.6 | 6.5 | 7 | 2 | {5.3.2.3 | 6.8 | 4.1.3 | 5.6.7 | 3 | 7 | | | | | | | | | | | {5.5.4
| | | | | | | | | | | | | | | | | | | | | | | | | " | 151- 160 | 8 | 4.3 | 6 | 5 | 6.7 | 2
| 4 | 5.6.7.8 | 4.3 | 5.6.8.7 | 5 | 5 | 8 |12 | 1: 1.50 5 | 161- 170 | 8 | 4.3 | 7.6
| 6.5 | 6.8.7 | 3.2 | 5.3.2.3.5.4 | 8 | 4.3 | 6.7 | 2 | 8 | | | " | 171- 176 | 8 |
2.4.3 | 7.6 | 6.5* | 8.7 | 3.2* | | | | | 1 | 5 | 3 |13 | 1: 4.33 7 | 177- 180 | | |
| | | | 5.4 | 8 | 4.4.3 | 8.7 | 1 | 3 | | | " | 181- 190 | 8 | 4.3 | 7.6 | 6.5 | 7 |
3.2 | 5.3.2.5.4 | 8 | 4.3 | 7 | 4 | 6 | 5 | 9 | 1: 1.80 8 | 191- 200 | 8 | 4.3 | 7.6
| 6.5 | 8.7 | 3.2 | 5.4 | 8 | 4.3 | 8.7 | 2 | 8 | | | " | 201- 210 | 8 | 4.3 | 6 |
6.4.6.5 | 8.7 | 3.2 | 5.4 | 8 | 4.3 | 7 | 4 | 6 | 6 |14 | 1: 2.33 9 | 211- 220 | 8 |
4.3 | 7.6 | 6.5 | 8.7 | 3.2 | 5.4 | 8 | 4.3 | 8.6.7 | 2 | 8 | | | " | 221- 230 | 9.8 |
4.3 | 7.6 | 6.5 | 6.7 | 3.2 | 5.4 | 7.6.8 | 4.3 | 7 | 1 | 9 | 3 |17 | 1: 5.67 | | | |
| | | | | | | | | | | 10 | 231- 240 | 8 | 4.3 | 7.6 | 6.5 | 6.7 | 3.2 | 5.4 |
{3.2.3.2.4.3 | 4.3 | 7 | 2 | 8 | 2 | 8 | 1: 4.00 | | | | | | | | | {2.5.4.7.8 | | | | |
| | | | | | | | | | | | | | | | | | | 11 | 241- 250 | 8 | 4.3 | 7.6 | 6.6.5 | 8.7 | 3.2 |
5.4 | 8 | 4.3 | 8.7 | 2 | 8 | 2 | 8 | 1: 4.00 | | | | | | | | | | | | | | | | | | 12 | 251-
260 | 8 | 4.3 | 6 | 6.5 | 6.7 | 3.3.2 | 5.4 | {7.6.7.7 | 3 | 3.7* | 3 | 7 | 3 | 7 | 1:
2.33 | | | | | | | | | {6.9.8* | | | | | | | | | | | | | | | | | | | | | | | 14 | 261-
270 | 8 | 3 | 6 | 6.5 | 7 | 3.2 | 5.4 | {5.3.4.3 | 3 | {3.3.3.3.4 | 5 | 5 | 5 | 5 | 1:
1.00 | | | | | | | | | {9.8* | | {4.6.4.7* | | | | | | | | | | | | | | | | | | | | | | | 15
| 271- 280 | 7.9.8 | 4.2.3 | 3.4.3.7.6 | 6.5 | 8.7 | 3.2 | 5.4 | 8 | 4.3 | 8.7 | 1 |
9 | | | " | 281- 290 | 8 | 4.3 | 7.6 | 6.5 | 7 | 3.2 | 5.4 | 7.8 | 4.3 | 7 | 3 | 7 | 4
|16 | 1: 4.00 | | | | | | | | | | | | | | | | | | | | | | | | | {4.3.2.3 | | | | | | | |
16 | 291- 300 | 7.8 | {4.4.4 | 6 | 6.5 | 7 | 3.3.2 | 5.4 | {6.5.4.3 | 4.3 | 6.7 | 2 |
8 | | | | | | {4.4.3 | | | | | {5.6.7.8 | | | | | | | | | | | | | | | | | | | | | | | "
| 301- 310 | 8 | 4.4.3 | 7.6 | 6.5 | 7 | 3.3.2 | 5.5.4 | {7.6.5.4.6 | 4.3 | 7 | 3 | 7
| 5 |15 | 1: 3.00 | | | | | | | | | {5.7.9.8 | | | | | | | | | | | | | | | | | | | | | | |
| 17 | 311- 320 | 7.8 | 4.3 | 7.6 | 6.6.5 | 7 | 3.2 | 5.4 | 7.6.7.6.7.8 | 4.3 | 7 | 2
| 8 | | | " | 321- 330 | 8 | 4.3 | 7.6 | 6.5 | 8.7 | 3.2 | 5.4 | 8 | 4.3 | 6.7 | 2 | 8
| 4 |16 | 1: 4.00 18 | 331- 340 | 7.7.8 | 4.3 | 6 | 6.5 | 7 | 3.2 | 5.4 | 8 | 4.3 |
7 | 4 | 6 | | | " | 341- 350 | 8 | 4.3 | 7.6 | 6.6.5 | 8.7 | 3.2 | 5.4 | 8 | 4.3 | 7 |
3 | 7 | 7 |13 | 1: 1.86 19 | 351- 360 | 8 | 4.3 | 7.6 | 6.5 | 6.5.6.5.7 | 3.2 | 5.4
| 8 | 4.3 | 7 | 3 | 7 | | | " | 361- 370 | 8 | 4.3 | 7.6 | 6.4.3.6.5 | 7 | 3.2 | 5.4 |

9.8 | 4.3 | 7 | 3 | 7 | 6 |14 | 1: 2.33 21 | 371- 380 | 8 | 4.3 | 7.6 | 6.5 | 7 |
3.2 | 5.4 | 7.8 | 4.3 | 8.7 | 2 | 8 | | | " | 381- 390 | 8 | 4.3 | 7.6 | 6.5 | 7 | 3.2
| 5.4 | 8 | 4.3 | 7 | 4 | 6 | 6 |14 | 1: 2.33 22 | 391- 400 | 8 | 4.3 | 7.6 | 6.5 |
6.5.4.6.7 | 3.3.3.2 | 5.4 | 6.7.8 | 4.4.3 | 7 | 2 | 8 | | | " | 401- 410 | 8 | 3 |
7.6 | 6.5 | 8.7 | 2 | 5.4 | 6.7.7.8 | 3 | 7 | 5 | 5 | 7 |13 | 1: 1.86 | | | | | | | | |
| | | | | | | 23 | 411- 420 | 8 | 4.4.3 | 7.6 | 6.5 | 8.7 | 3.2 | 5.4 | {7.6.7.6 |
4.3 | 7 | 2 | 8 | 2 | 8 | 1: 4.00 | | | | | | | | | {6.7.8 | | | | | | | | | | | | | | | |
| | | | | | | | 24 | 421- 430 | 8 | 4.3 | 6 | 6.5 | 7 | 3.2 | 5.4 | 8 | 3 | 7 | 6 | 4
| | | " | 431- 440 | 8 | 3 | 7.6 | 6.5 | 7 | 3.3.2 | 5.4 | 8 | 4.3 | 7 | 5 | 5 |11 | 9
| 1: 0.82 25 | 441- 450 | 7.8 | 4.4.3 | 7.6 | 6.5 | 6.5.7 | 3.3.2 | 5.5.4 | 7.8 |
4.3 | 7 | 1 | 9 | 1 | 9 | 1: 9.00 26 | 451- 460 | 7.8 | 4.3 | 7.6 | 6.5 | 7 | 3.2 |
5.4 | 8 | 4.3 | 7 | 3 | 7 | | | " | 461- 470 | 8 | 4.3 | 7.6 | 6.5 | 7 | 3.2 | 5.4 | 8
| 4.3 | 8.7 | 3 | 7 | 6 |14 | 1: 2.33 28 | 471- 480 | 8 | 4.4.3 | 7.6 | 6.5 | 7 |
3.2 | 5.4 | 9.8 | 3 | 8.7 | 3 | 7 | 3 | 7 | 1: 2.33 29 | 481- 490 | 8 | 4.3 | 7.7.6 |
6.5 | 7 | 3.2 | 5.4 | 8 | 4.3 | 8.7 | 3 | 7 | 3 | 7 | 1: 2.33 30 | 491- 500 | 7.9.8 |
4.3 | 7.6 | 6.5 | 7 | 3.2 | 5.4 | 8 | 4.3 | 7 | 3 | 7 | | | " | 501- 510 | 8 | 4.3 |
7.6 | 6.5 | 8.7 | 3.2 | 5.4 | 8 | 4.4.3 | 8.7 | 2 | 8 | 5 |15 | 1: 3.00 July | | | | |
| | | | | | | | | | | 1 | 511- 520 | 8 | 4.3 | 7.6 | 6.5 | 7 | 3.2 | 5.4 | 8 | 4.3 | 7
| 4 | 6 | 4 | 6 | 1: 1.50 | | | | | | | | | | | | | | | | | 2 | 521- 530 | 8 | 4.3 | 7.6
| 6.5 | 7 | 3.2 | 5.4 | {7.6.5.6.5 | 4.4.3 | 7 | 3 | 7 | | | | | | | | | | | | | {6.5.6.8
| | | | | | | | | | | | | | | | | | | | | | | | " | 531- 540 | 8 | 4.3 | 7.6 | 6.5 | 7 |
3.2 | 5.4 | 8 | 4.3 | 7 | 4 | 6 | 7 |13 | 1: 1.86 3 | 541- 550 | 7.8 | 4.4.3 | 6 |
6.5 | 7 | 3.2 | 5.4 | 8 | 4.3 | 5.5.7 | 3 | 7 | | | " | 551- 560 | 7.8 | 4.3 | 6 | 6.5
| 7 | 3.2 | 5.4 | 8 | 3 | 7 | 5 | 5 | 8 |12 | 1: 1.50 5 | 561- 570 | 7.7.8 | 4.3 | 6
| 6.5 | 6.7 | 3.3.2 | 5.4 | 8 | 4.3 | 7 | 3 | 7 | | | " | 571- 580 | 8 | 4.3 | 6 | 6.5
| 7 | 3.2 | 5.4 | 7.8 | 4.3 | 7 | 4 | 6 | 7 |13 | 1: 1.86 | | | | | | | | | | | | | | |
| | | | | | | {6.5.4.6.5 | | | | | | | | | | 6 | 581- 590 | 7.8 | 4.3 | 7.6 | 6.6.5 |
{5.4.5.4.4 | 2 | 3.4 | 6.5.4.3.7.8 | 3 | 7 | 3 | 7 | 3 | 7 | 1: 2.33 | | | | | | |
{6.5.6.5.8.7 | | | | | | | | | | | | | | | | | | | | | | | | | | | | | 7 | 591- 600 | 8 | 4.3
| 6 | 6.5 | 7 | 3.2 | 5.4 | 7.8 | 4.3 | 8.7 | 3 | 7 | | | " | 601- 610 | 7.8 | 4.3 |
7.6 | 6.5 | 8.7 | 3.2 | 5.4 | 8 | 4.3 | 8.7 | 1 | 9 | 4 |16 | 1: 4.00 8 | 611- 620 |
8 | 4.3 | 7.6 | 6.5 | 8.7 | 3.2 | 5.4 | 8 | 4.3 | 8.7 | 2 | 8 | | | " | 621- 630 | 8 |
4.3 | 7.6 | 6.5 | 8.7 | 3.2 | 5.4 | 9.8 | 4.3 | 8.7 | 1 | 9 | | | " | 631- 640 | 8 |
4.4.3 | 7.7.6 | 6.5 | 8.7 | 3.2 | 4 | 8 | 4.3 | 8.7 | 3 | 7 | 6 |24 | 1: 4.00 | | | |
| | | | | | | | | | 9 | 641- 650 | 7.8 | 4.3 | 7.6 | 6.5 | 6.7 | 3.2 | {3.2.5.3 |
7.6.5.4.8 | 3 | 8.7 | 1 | 9 | 1 | 9 | 1: 9.00 | | | | | | | | {2.5.4 | | | | | | | | |
| | | | | | | | | | | | | | | | | | | | | | {6.5.4.3.7 | | | | | | | 10 | 651- 660 |
7.8 | 4.3 | 6 | 6.5 | 7 | 3.2 | 5.4 | {6.5.4.7.6 | 4.3 | 7 | 3 | 7 | | | | | | | | | | |

| {5.4.8 | | | | | | | | | | | | | | | | | | | | | | | 10 | 661- 670 | 8 | 3 | 7.6 | 5 |
7 | 3.2 | 5.4 | 8 | 3 | 8.7 | 6 | 4 | 9 |11 | 1: 1.22 12 | 671- 680 | 7.8 | 3 | 6 |
6.5 | 7 | 3.2 | 5.4 | 6.5.4.7.8 | 4.3 | 8.7 | 3 | 7 | 3 | 7 | 1: 2.33 | | | | | | | | |
| | | | | | | 13 | 681- 690 | 8 | 3 | 7.6 | 6.5 | {6.5.4 | 3.2 | 4 | 6.7.8 | 3 |
{6.5.4.5 | 4 | 6 | | | | | | | | | | {6.5.7 | | | | {6.5.8.7 | | | | | | | | | | | | |
| | | | | | | " | 691- 700 | 8 | 3 | 6 | 5 | 7 | 3.2 | 5.4 | 8 | 3 | 7 | 8 | 2 |12 | 8
| 1: 0.67 | | | | | | | | | | | | | | | | | 14 | 701- 710 | 8 | 3 | 7.6 | 6.5 | {6.5.4.5
| 2 | 3.5.4 | 8 | 3 | 7 | 6 | 4 | | | | | | | | {4.6.8.7 | | | | | | | | | | " | 711-
720 | 8 | 3 | 6 | 5 | 7 | 2 | 5.4 | 6.5.4.8 | 3 | 6.5.7 | 7 | 3 |13 | 7 | 1: 0.54 15
| 721- 730 | 7.8 | 3 | 6 | 6.5 | 7 | 3.2 | 5.4 | 8 | 4.3 | 7 | 5 | 5 | | | " | 731-
740 | 8 | 3 | 7.6 | 6.5 | 7 | 3.2 | 5.4 | 8 | 3 | 7 | 6 | 4 |11 | 9 | 1: 0.82 16 |
741- 750 | 7.8 | 3 | 6 | 6.5 | 7 | 3.2 | 4 | 8 | 3 | 7 | 7 | 3 | | | " | 751- 760 |
7.8 | 3 | 7.6 | 6.5 | 7 | 2 | 4 | 7.8 | 4.3 | 7 | 5 | 5 |12 | 8 | 1: 0.67 17 | 761-
770 | 8 | 4.3 | 6 | 5 | 8.7 | 3.2 | 5.4 | 8 | 4 | 7 | 6 | 4 | | | " | 771- 780 | 8 |
2.2.3 | 7.6 | 6.5 | 7 | 3.2 | 5.4 | 7.8 | 4.3 | 7 | 3 | 7 | 9 |11 | 1: 1.22 19 | 781-
790 | 8 | 3 | 7.6 | 5 | 7 | 3.2 | 3.4 | 7.6.5.8 | 3 | 7 | 6 | 4 | | | " | 791- 800 |
7.8 | 3 | 6 | 5 | 7 | 2 | 5.4 | 8 | 3 | 6.5.6.7 | 7 | 3 | | | " | 801- 810 | 8 | 2.3 |
6 | 5 | 6.5.7 | 2 | 5.4 | 8 | 3 | 7 | 7 | 3 |20 |10 | 1: 0.50 20 | 811- 820 | 7.8 |
3 | 7.6 | 5 | 7 | 3.3.2 | 5.4 | 8 | 2.2.3 | 7 | 5 | 5 | | | " | 821- 830 | 8 | 3 | 6 |
6.5 | 7 | 3.2 | 4 | 8 | 2.3 | 8.7 | 6 | 4 |11 | 9 | 1: 0.82 21 | 831- 840 | 8 | 3 |
5.4.5.6 | 5 | 7 | 2 | 4 | 6.7.8 | 3 | 8.7 | 7 | 3 | | | " | 841- 850 | 8 | 3 | 7.6 | 5
| 7 | 3.2 | 3.2.4 | 8 | 3 | 7 | 7 | 3 |14 | 6 | 1: 0.43 22 | 851- 860 | 8 | 4.3 | 6 |
5 | 7 | 3.2 | 3.5.4 | 8 | 3 | 8.7 | 6 | 4 | | | " | 861- 870 | 7.8 | 4.3 | 7.6 | 6.5 |
8.7 | 3.2 | 4 | 8 | 3 | 8.7 | 3 | 7 | | | " | 871- 880 | 8 | 4.3 | 7.6 | 6.5 | 8.7 |
3.2 | 5.4 | 8 | 4.3 | 8.7 | 2 | 8 |11 |19 | 1: 1.73 23 | 881- 890 | 8 | 3 | 7.6 |
6.5 | 8.7 | 3.2 | 5.4 | 8 | 3 | 8.7 | 4 | 6 | | | " | 891- 900 | 8 | 3 | 7.6 | 6.5 | 7
| 3.3.2 | 5.4 | 8 | 4.3 | 7 | 5 | 5 | 9 |11 | 1: 1.22 24 | 901- 910 | 8 | 4.3 | 7.6
| 6.5 | 7 | 3.2 | 5.4 | 8 | 3 | 8.7 | 4 | 6 | | | " | 911- 920 | 8 | 3 | 7.6 | 5 | 7 |
3.2 | 5.4 | 8 | 3 | 7 | 7 | 3 |11 | 9 | 1: 0.82 26 | 921- 930 | 7.8 | 3 | 7.6 | 5 |
7 | 3.2 | 5.4 | 8 | 2.2.3 | 8.7 | 4 | 6 | | | " | 931- 940 | 8 | 3 | 7.6 | 6.5 | 8.7 |
3.2 | 5.5.4 | 8 | 4.3 | 8.7 | 3 | 7 | 7 |13 | 1: 1.86 27 | 941- 950 | 8 | 3 | 6 |
6.5 | 7 | 3.2 | 5.4 | 8 | 3 | 7 | 7 | 3 | 7 | 3 | 1: 0.43 28 | 951- 960 | 8 | 3 | 7.6
| 6.5 | 5.4.7 | 2 | 5.5.4 | 8 | 3 | 7 | 6 | 4 | 6 | 4 | 1: 0.67 29 | 961- 970 | 8 | 3
| 7.6 | 5 | 8.7 | 2 | 4 | 8 | 3 | 7 | 8 | 2 | 8 | 2 | 1: 0.25 | | | | | | | | | | | | | |
| | 30 | 971- 980 | 8 | 3 | 4.3.2.6 | 5 | {6.5.4.6 | 2 | 5.5.4 | 8 | 3 | 7 | 7 | 3 |
7 | 3 | 1: 0.43 | | | | | {6.5.7 | | | | | | | | | | | | | | | | | | | | | | | | | | | | | 31
| 981- 990 | 8 | 3 | 6 | 6.5 | 8.7 | 2 | 4 | 8 | 3 | 7 | 8 | 2 | 8 | 2 | 1: 0.25 |
August | | | | | | | | | | | | | | | | | | 2 | 991-1000 | 8 | 3 | 7.6 | 5 | 7 | 2 |

{2.3.5.3 | 7.6.8 | 3 | 7 | 7 | 3 | 7 | 3 | 1: 0.43 | | | | | | | | {2.3.3.4 | | | | | |
| | | | | | | | | | | | | | | | | | | | 3 | 1001-1010 | 8 | 3 | 7.6 | 5 | 7 | 2 | 4 |
7.6.5.6.7.8 | 3 | 5.4.3.4.3.7 | 7 | 3 | | | " | 1011-1020 | 8 | 2.3 | 5.6 |
3.2.3.6.5 | 7 | 2 | 5.4 | 9.8 | 2.1.3 | 7 | 4 | 6 |11 | 9 | 1: 0.82 | | | | | | | | | |
| | | | | | 4 | 1021-1030 | 7.8 | 3 | 5.4.3.7.6 | 6.5 | {6.5.6 | 3.2 | 5.4 | 8 |
{2.2.4.2 | 8.7 | 2 | 8 | | | | | | | | | {5.6.7 | | | | {4.2.3 | | | | | | | | | | | | |
| | | | | | | | | " | 1031-1040 | 7.8 | 3 | 6 | 6.4.3.6.5 | 7 | 2 | 3.5.4 | 8 | 2.3 |
8.7 | 5 | 5 | 7 |13 | 1: 1.86 | | | | | | | | | | | | | | | | 5 | 1041-1050 | 8 | 3 |
6 | 2.3.2.6.5 | 8.7 | 2 | 4 | 8 | 2.2.4.3 | {8.8.6.8.4 | 6 | 4 | 6 | 4 | 1: 0.67 | | |
| | | | | | | | {6.5.8.7 | | | | | | | | | | | | | | | | | | | | | 6 | 1051-1060 | 8 |
3 | 6 | 4.2.6.5 | 7 | 3.2 | 5.4 | 8 | 3 | 8.7 | 6 | 4 | 6 | 4 | 1: 0.67 | | | | | | | |
| | | | | | | | 7 | 1061-1070 | 8 | 3 | 5.4.3.6 | 4.5 | {6.5.6.5 | 2 | 4 | 8 | 3 | 7
| 7 | 3 | 7 | 3 | 1: 0.43 | | | | | | {4.8.7 | | | | | | | | | | | | | | | | | | | | | | | |
| | | 9 | 1071-1080 | 8 | 3 | 6 | 5 | 7 | 2 | 4 | 8 | 3 | 7 |10 | 0 |10 | 0 | 1:
0.00 --------+------------+--------------+---------------+---------------+---------------+------------
--+--------------+---------------+---------------+---------------+---------------+---+---+---+---+---
----- | | | | 1.2.3.4.5 | | | | | | 1.2.3.4 | 2.3.4.5 | | | | | | | 5.6.7.8 | 2.3.4.5.6
| 6.7.8.9 | 5.6.7 | 1.2.3.4 | 4.5.6 | 2.3.4.5 | 1.2.3 | 5.6.7 | 6.7.8.9 | | | | | -----
---+------------+--------------+---------------+---------------+---------------+--------
------+--------------+--------------+---------------+---+---+---+---+--------- 10 |
1- 10 | 6.5.7 | 3.2.6.5 | 8 | 6 | 2.4.3 | 5 | 5.4 | 2 | 7.5.2.7.6 | 8 | 5 | 5 | 5 | 5
| 1: 1.00 11 | 11- 20 | 7 | 3.6.5 | 8 | 6 | 3 | 6.5 | 4 | 3.2 | 7.6 | 8 | 6 | 4 | 6 |
4 | 1: 0.67 | | | | | | | | | | | | | | | | | | | | | | | | {3.2.3.5.3 | | | | | | | | 12
| 21- 30 | 7 | 2.2.6.5 | 7.8 | 6 | 3 | 5 | {2.5.3.2 | 2 | 6 | 8 | 7 | 3 | 7 | 3 | 1:
0.43 | | | | | | | | {5.3.2.5 | | | | | | | | | | | | | | | | {2.5.5.4 | | | | | | | | |
| | | | | | | | | | | | | | | |
========+===========+==============+===============+============
==+==============+===============+===============+===============+
==============+==============+==============+===+===+===+===+==
======

[Footnote 1: First choices correct by reason of aid from the experimenter
are not counted as correct (R) in the summary.]

[Footnote *: Aided by experimenter.]

The method of choosing the first box at the right end and then the one next
to it developed in the case of all except two of the ten settings. The time of

appearance is worth noting. In setting 1, it failed to appear; in setting 2, it developed early,--after about one hundred trials; in setting 3, after about one hundred and fifty trials; in setting 4, after about one hundred and fifty trials; in setting 5, after about one hundred and seventy trials; in setting 6, after about one hundred trials; in setting 7, after about fifty trials; in setting 8, it never developed; in setting 9, after about fifty trials; and in setting 10, it developed very late,--after about four hundred and seventy trials.

This method of reaction, although inadequate, proved remarkably persistent, and it is doubtful whether it had been wholly overcome at the conclusion of the experiment. In the case of the series of trials given on June 8, numbered 191 to 200, the method used was either that of the first at the right and then the next, or direct choice of the right box.

Throughout the trials with this problem, the end boxes, numbers 1 and 9, were avoided. This is at least partially explained by the fact that they never existed, and obviously never could appear, in problem 2, as right boxes. In trials 601 to 610, given on July 7, there occurred partial return to the formerly established method of choosing the first door at the right. This relapse was characteristic of what happened during the many days which intervened between the definite appearance of this habit and the final solution of the problem.

Especially in connection with such relapses, Skirrl showed extreme fatigue or ennui and often would refuse to work and simply sit before the open doors yawning. This happened even when he was extremely hungry and evidently eager enough for food.

From July 12 on the hunger motive was increased by feeding the monkey only in the apparatus and by so regulating the amount of food given in each trial that he should obtain barely enough to keep him in good physical condition. An increase in the number of correct choices promptly resulted, and continued until on July 14 the ratio of choices was 1 to .54. It appeared from these data that a relatively small number of choices, say not more than ten a day, the rewards in connection with which supplied the only food received by the animal, yielded most favorable results.

On July 16, the period of confinement in wrong boxes was increased to sixty seconds, and it was so continued for a number of days. But in the end, it became clear that the period of thirty seconds, combined with a liberal reward in the shape of desired food and a single series of ten trials per day, was most satisfactory. The detailed data of table 2 indicate that at this time Skirrl was making his choices by memory of the particular setting.

Skirrl, on July 17 was evidently hungry and eager to locate food, but seemingly unable to select the right box. In trial 5 (765th) of the series, he was punished by confinement in box 8. When the doors were unlocked in order that the entrance door might be raised to release him, the lock-bar, sliding under the floor, made a slight grating noise, and the instant the entrance door was opened, he jumped out excitedly. _He made no outcry, but as soon as he was out of the box, sat down, and taking up his right hind foot, examined it for a few seconds._ Having apparently assured himself that nothing serious had happened, he went on unconcernedly about his task. The presumption is that the sound of the lock-bar, associated as it was with his painful experience in box 1, revived the strongly affective experience of stepping on the nail. Psychologically described, the sound induced an imaginal complex equivalent to the earlier painful experience. The behavior seems to the writer a most important bit of evidence of imagery in the monkey. Finally, on August 9, after ten hundred and seventy trials, Skirrl succeeded in choosing correctly in the ten trials of a series, and he was therefore considered to have solved the problem of the second door from the right end of the group.

On the following day, he was given a control series with the settings which are presented on page 19 and also at the bottom of table 2. In this series he chose correctly five times,--in other words, as often correctly as incorrectly. An analysis of the choices indicates, however, that two of the five correct choices were made in box 8, which, as it happened, had proved a peculiarly easy one for him throughout the training, since from the first he tended to avoid door 9. Consequently, it is only fair to conclude, from the results for this control series and for those given on August 11 and 12, that the animal chose not on the basis of anything remotely resembling a general idea of secondness from the right end, but instead on the basis of gradually acquired modes of reaction to the particular settings. This conclusion is strengthened by the fact that he had failed to learn to react appropriately and readily to most of the settings of the regular series.

The curve which represents the course of the learning process in this problem is presented in figure 19. For this and all other curves which involve more than a single series of observations a day, the method of construction was as follows: The first series for each day of training is indicated on the curve by a dot, while the second or third series on a given day, although space is allowed for them, are not so indicated. Consequently, the form of the curve is determined chiefly by the first series per day. The extreme irregularities of this curve are most interesting and puzzling, as are also the variations in the daily ratios of right to wrong first choices. Three times in the course of the training, this ratio rose to 1 to 9, or higher. The causes for such extreme variations are not easily enumerated, but a few of the most obvious contributory causes are variations in the weather, especially cloudiness or fogginess, which rendered the apparatus dark; variations in the degree of hunger or eagerness for food; differences in the activities of the animals in the cages outside of the laboratory (sometimes they were noisy and distracted the subject), and finally, differences in the physical fitness and attitude of the animal from day to day.

The more or less incidental behavior in connection with this experiment more strongly than the statistical results of the work on problem 2 indicate the existence of imagery. That ideas played a part in the solution of the problem is probable, but at best they functioned very ineffectively. The small number of methods used in the selection of the right box, and the slight variations from the chief method, that of choosing the first box at the right end and then the one next to it, apparently justify Doctor Hamilton's characterization of this monkey as defective.

Problem 3. Alternately First at Left and First at Right_

Following the control series given in connection with problem 1, an interval of rest lasting from August 12 to August 19 was allowed in order that Skirrl might in part at least lose the effects of his training and regain his customary interest in the apparatus by being allowed to obtain food easily instead of by dint of hard labor,--labor which was harder by far, apparently, than physical activity because it demanded of the animal certain mental processes which were either lacking or but imperfectly functional. The difficultness of the daily tasks appears to be reliably indicated by the tendency to yawn.

Systematic work on problem 3, which has been defined as alternately the first door at the left and the first door at the right of the group, was begun August 19, and for nine days a single series of ten trials per day was given. Work then had to cease because of the experimenter's return to Cambridge.

The results of the work on this problem demand but brief analysis and comment. The expected ratio of one right to four wrong choices per series appears (see table 3) for the first series of trials, and this in spite of the fact that Skirrl had been trained for several weeks to choose the second door from the right end. One would ordinarily have predicted a much larger number of incorrect choices. The right choices were due to the monkey's strong tendency to go first to the first door at the right and thence to the one next to it. Indeed in the series given on August 24; this method was followed without variation. In other words, in every one of the ten trials Skirrl entered first the box at the extreme right end of the group. This necessarily resulted in as many right as wrong first choices. Consequently, the ratio reads 1 to 1. But the method was not adhered to, and at no time either before or after that date did he succeed in equalling this achievement. There was, as a matter of fact, no steady improvement, and so far as one may judge from the records which were obtained, the course of events in the solution of this problem would have been similar to those in problem 2.

TABLE 3

Results for Skirrl, _P. irus_, in Problem 3

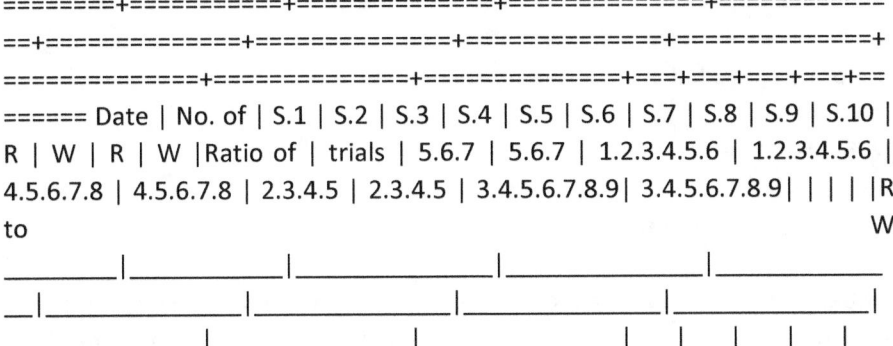

```
========+===========+==============+==============+===========
==+=============+==============+==============+==============+
=============+==============+==============+===+===+===+===+==
====== Date | No. of | S.1 | S.2 | S.3 | S.4 | S.5 | S.6 | S.7 | S.8 | S.9 | S.10 |
R | W | R | W |Ratio of | trials | 5.6.7 | 5.6.7 | 1.2.3.4.5.6 | 1.2.3.4.5.6 |
4.5.6.7.8 | 4.5.6.7.8 | 2.3.4.5 | 2.3.4.5 | 3.4.5.6.7.8.9| 3.4.5.6.7.8.9| | | | |R
to                                                                              W
_____|_____|_____|_____|_____
__|_____|_____|_____|_____|
_____|_____|_____|__|__|__|__|__
```

August | | | | | | | | | | | | | | | | | | | | | {6.5.4.6 | | | | | | | {8.7.6.5 |
| | | | 19 | 1- 10 | 7.5 | 6.7 | {3.2.6.4 | 5.3.6 | {7.8.7.6 | 8 | 4.3.5.4.5.2 | 5 |
9.8.7.6.4.3 | {4.5.8.7 | 2 | 8 | 2 | 8 | 1:4.00 | | | | {3.6.1* | | {8.6.4 | | | | |
{6.5.9* | | | | | | | | | | | | | | | | | | | | | | | | | | | | | {5.4.6.5 | | {8.7.6.7 | | | | |
| | | | | 20 | 11- 20 | 7.6.5 | {6.5.6.5 | {4.6.5.3 | 5.3.2.4.6 | {7.5.8.8 | 8 |
5.4.3.2 | 4.5 | 8.7.6.5.4.3 | {8.7.6.8.3 | 1 | 9 | 1 | 9 | 1:9.00 | | | {7 | {2.5.1* |
| {7.6.4* | | | | | {7.6.4.3.9 | | | | | | | | | | | | | | | | | | | | | | | | | | {6.2.5.6
| | | | {5.3.5.4 | | | | | | | | 21 | 21- 30 | 7.6.7.6.5 | 7 | {5.3.6.5 | 2.5.5.6 |
8.6.5.4 | 8 | {3.5.3.5 | 5 | {9.8.7.6 | 8.9 | 3 | 7 | 3 | 7 | 1:2.33 | | | | {4.3.1* |
| | | {4.3.2* | | {5.4.3 | | | | | | | | | | | | | | | | | | | | | | | | | | | | | | | | |
| | | | | 23 | 31- 40 | 7.6.5 | 6.5.7 | {6.4.3.2 | 3.2.6 | 8.7.6.4 | 8 | 5.4.5.3.2 |
5.2 | 8.7.3 | 9 | 2 | 8 | 2 | 8 | 1:4.00 | | | | {5.6.2.1 | | | | | | | | | | | | | |
| | | | | | | | | | | | | | | | | | | | | | | | | | | | | 24 | 41- 50 | 7.6.5 | 7 |
{6.2.5 | 6 | {8.7.8.7 | 8 | 5.3.2 | 5 | {9.8.7.6 | 9 | 5 | 5 | 5 | 5 | 1:1.00 | | | |
{4.3.1 | | {5.7.5.4 | | | | {5.4.3 | | | | | | | | | | | | | | | | | | | | | | | | |
| | | | | {8.7.3.6 | | | | | 25 | 51- 60 | 7.6.5 | 6.5.7 | 5.2.1 | 6 | 8.5.4 | 8 | 2 |
2.5 | 9.8.7.4.3 | {8.7.5.3 | 3 | 7 | 3 | 7 | 1:2.33 | | | | | | | | | | | {8.7.9* | | |
| | | | | | | | | | | | | | | | | | | | | | | | | | | | | | 26 | 61- 70 | 7.6.5 |
6.5.7 | 1 | 2.1.6 | 8.7.6.4 | 8 | 2 | 3.2.5 | 9.8.7.5.3 | {3.6.8.3 | 3 | 7 | 3 | 7 |
1:2.33 | | | | | | | | | | | {4.7.9 | | | | | | | | | | | | | | | | | | | | | | | |
| | | | | {8.6.3.3 | | | | | 27 | 71- 80 | 7.6.5 | 7 | 2.1 | 1.5.6 | 8.7.6.4 | 8 | 2 |
5 | 9.8.7.6.3 | {7.5.3.8 | 4 | 6 | 4 | 6 | 1:1.50 | | | | | | | | | | | {3.6.9* | | | |
| | | | | | | | | | | | | | | | | | | | | | | | | | | | | | 28 | 81- 90 | 7.6.7.5 |
7 | 3.1 | 6 | 8.6.4 | 4.8 | 2 | 2.4.5 | 8.7.4.3 | 3.8.9 | 3 | 7 | 3 | 7 | 1:2.33 | | |
| | | | | | | | | | | | | |
=========+============+===============+===============+============
==+===============+===============+===============+===============+
==============+===============+===============+===+===+===+===+==
======

[Footnote *: Aided by experimenter.]

2. Sobke, Pithecus rhesus

Problem 1. First at the Left End

Sobke was somewhat afraid of the experimenter when the investigation was undertaken, and instead of willingly coming out of his cage when the door was raised, he often had to be coaxed out and lured into the apparatus with food. Whereas Skirrl was frank and rather aggressive, Sobke was stealthy in his movements, furtive, and evidently suspicious of the experimenter as well as of the apparatus. He was perfectly safe to approach, but would not permit anyone to touch him. After a few days, he began to take food from the hands of the experimenter.

Preliminary work to acquaint this monkey with the routine of the experiment was begun on April 13. As in the case of Skirrl, he was lured into the apparatus and was taught the route through the boxes to the starting point by being allowed to obtain food once each day in each of the nine boxes. The procedure was simple. The entrance door and the exit door of a particular box were raised and the animal admitted to the reaction-compartment and permitted to pass through the box whose doors stood open, take its food, and return to the starting point. Sobke very quickly learned the route perfectly and came to work steadily and rapidly. After five days of preliminary work of this sort, he was so thoroughly accustomed to the apparatus that it was evidently desirable to begin with regular training experiments.

The first series of trials was given on April 19. Both punishment and reward were employed from the first. The punishment consisted of confinement for thirty seconds in each wrong box, and the reward of a small piece of banana, usually not more than a tenth of a medium sized banana for each correct choice. The total time for the first series of trials was fourteen minutes. This indicates that Sobke worked rapidly. My notes record that he worked quickly though shyly, wasted almost no time, made few errors of choice, and waited quietly during confinement in the boxes. In this, also, he differed radically from Skirrl who was restless and always tried to escape from confinement.

Throughout the work on problem 1, punishment and reward were kept constant. Everything progressed smoothly; there were no such irregularities of behavior as appeared in the case of Skirrl, and consequently the description of results is a relatively simple matter. Sobke invariably chose the end boxes. His performance was in every way superior to that of Skirrl.

As previously, the detailed results are presented in tabular form (table 4). From this table it appears that, whereas the expected ratio of right to wrong first choices for this problem is 1 to 2.5, the actual ratio for Sobke's first series was 1 to .67. This surprisingly good showing is unquestionably due to his marked tendency to choose the end box of a group; and this tendency, in turn, may in part be the result of the preliminary training, for during that only one box was open each time. But, if the preliminary training were responsible for Sobke's tendency, it should be noted that it had very different effect upon Skirrl, and, as will be seen later, upon Julius.

The results for the ten different settings of the doors for problem 1 as they appear in table 4 are of interest for a number of reasons. In the first place, the setting 1. 2. 3 appearing twice,--at the beginning of the series and again at the end--yielded markedly different results in the two positions. For whereas no mistakes were made in the case of setting 1, there were fifty per cent of incorrect first choices for setting 10. Again, satisfactory explanation is impossible. It is conceivable that fatigue or approaching satiety may have had something to do with the failures at the end of the series, but as a rule, as is indicated by settings 1, 2, and 6, if correct choices were made at the beginning, they continued throughout the day's work.

In this problem, Sobke's improvement was steady and fairly rapid, and in the eighth series, trials 71 to 80, only correct first choices appear. Consequently, seventy trials were required for the solution of the problem. This number is in marked contrast with Skirrl's one hundred and thirty-two trials.

Immediately following the first perfect series, Sobke was given two series of control tests on April 28. Conditions were unfavorable, since the day was stormy and the rain pattering on the sheet-iron roof made a great din. Nevertheless, he worked steadily and well up to the sixth trial, which was preceded by a slight delay because of the necessity of refilling some of the food boxes. After this interruption, wrong choices occurred in trial 6. And again after trial 9, there was brief interruption, followed by wrong choices in trial 10. The ratio of right to wrong choices for this first control series was therefore 1 to .25.

TABLE 4

Results for Sobke, _P. rhesus_, in Problem 1

========+===========+==============+==============+===========

```
==+=============+=============+=============+=============+
=============+=============+=============+===+===+===+===+==
====== | No. | S.1 | S.2 | S.3 | S.4 | S.5 | S.6 | S.7 | S.8 | S.9 | S.10 | | | | |
Ratio Date | of | | | | | | | | | | R | W | R | W | of | trials | 1.2.3 | 8.9 |
3.4.5.6.7 | 7.8.9 | 2.3.4.5.6 | 6.7.8 | 5.6.7 | 4.5.6.7.8 | 7.8.9 | 1.2.3 | | | | | R
to W --------+-----------+--------------+-------------+--------------+--------------+-----------
---+--------------+--------------+--------------+--------------+--------------+---+---+---+---+--
------ April | | | | | | | | | | | | | | | | | 19 | 1-10 | 1 | 8 | 3 | 9.7 | 6.2 | 6 | 7.5
| 4 | 9.7 | 1 | 6 | 4 | 6 | 4 | 1:0.67 20 | 11-20 | 1 | 8 | 3 | 7 | 2 | 6 | 7.5 | 8.4
| 9.9.7 | 1 | 7 | 3 | 7 | 3 | 1:0.43 21 | 21-30 | 1 | 8 | 4.3 | 9.7 | 2 | 6 | 5 | 8.4
| 7 | 1 | 7 | 3 | 7 | 3 | 1:0.43 22 | 31-40 | 1 | 8 | 3 | 7 | 6.2 | 6 | 6.5 | 4 | 7 |
3.1 | 7 | 3 | 7 | 3 | 1:0.43 23 | 41-50 | 1 | 8 | 3 | 7 | 2 | 6 | 5 | 4 | 9.7 | 3.1 |
8 | 2 | 8 | 2 | 1:0.25 24 | 51-60 | 1 | 8 | 3 | 9.7 | 2 | 6 | 5 | 4 | 7 | 2.1 | 8 | 2
| 8 | 2 | 1:0.25 26 | 61-70 | 1 | 8 | 3 | 7 | 2 | 6 | 5 | 4 | 7 | 3.1 | 9 | 1 | 9 | 1
| 1:0.11 27 | 71-80 | 1 | 8 | 3 | 7 | 2 | 6 | 5 | 4 | 7 | 1 |10 | 0 |10 | 0 | 1:0.00
--------+-----------+--------------+--------------+--------------+--------------+--------------+---
-----------+--------------+--------------+--------------+--------------+---+---+---+---+--------
| | | | | | | | 2.3.4 | | | 1.2.3.4.5 | | | | | | 2.3.4 | 6.7.8.9 | 3.4.5 |
4.5.6.7.8.9 | 6.7.8.9 | 1.2.3.4.5 | 5.6.7.8 | 3.4.5.6.7.8 | 5.6.7 | 6.7.8.9 | | | | |
| +--------------+--------------+--------------+--------------+--------------+--------------+-----
--------+--------------+--------------+--------------+ | | | | 28 | 1-10 | 2 | 6 | 3 | 4 | 6
| 5.4.1 | 2 | 3 | 5 | 5.4.2.1 | 8 | 2 | | | " | 11-20 | 2 | 6 | 3 | 4 | 6 | 2.1 | 2 | 3
| 5 | 1 | 9 | 1 |17 | 3 | 1:0.18 | | | | | | | | | | | | | | | | |
========+===========+=============+=============+=============
==+=============+=============+=============+=============+
=============+=============+=============+===+===+===+===+==
======
```

Six minutes after completion of the first control series, a second was given under slightly more favorable conditions, and in this only a single wrong choice occurred, in that box 2 was first chosen in trial 6 instead of box 1. From the results of these two control series, it is evident that Sobke's solution of problem 1 is reasonably adequate. He is easily diverted or disturbed in his work by any unusual circumstances, but so long as everything goes smoothly, he chooses with ease and certainty. Whether it is fair to describe the behavior as involving an idea of the relation of the right box to the other members of the group would be difficult to decide. I hesitate to infer definite ideation from the available evidence, but I strongly suspect the presence of images and relatively ineffective or inadequate ideation.

It is perfectly evident that Sobke is much more intelligent than Skirrl. In practically every respect, he adapted himself more quickly to the experimental procedure and progressed more steadily toward the solution of the problem than did Skirrl. The contrast in the learning processes of the two monkeys could scarcely be better exhibited than by the curves of learning which are presented in figure 18. The first, that for Sobke, is surprisingly regular; the second, that for Skirrl, is quite as surprisingly irregular. These results correlate perfectly with the steadiness and predictability of the former's responses and the irregularity and erraticness of the latter's.

Problem 2. Second from the Right End

On the completion of problem 1 Sobke was in perfect condition, as to health and training, for experimental work. He had come to work quietly, fairly deliberately, and very steadily. His timidity had diminished and he would readily come to the experimenter for food, although still he was somewhat distrustful at times and became timid when anything unusual occurred in the apparatus.

As preparation for problem 2, a break in regular experimentation covering four days followed the control series of problem 1. On each of these four days the monkey was allowed to get food once from each of the nine boxes, both doors of a given box being open for the trial and all other doors closed. For this feeding experiment, the doors were opened in irregular order, and this order was changed from day to day.

Systematic work with problem 2 began on May 3, with punishment of thirty seconds for mistakes and a liberal reward of food for each success. Early in the series of trials it was discovered that Sobke was likely to become discouraged and waste a great deal of time unless certain aid were given by the experimenter. On this account, after the first two trials, the method was adopted of punishing the animal by confinement for the first ten mistakes in a trial, and of then, if need be, indicating the right box by slightly and momentarily raising the exit door. Every trial in which aid was thus given by the experimenter is indicated in table 5 by an asterisk following the last choice. In the first series of trials for this problem, aid had to be given in seven of the ten trials, and even so the series occupied seventy-one minutes. It is possible that had no aid been given, the work might have been continued successfully with a smaller number of trials than ten per day. But under the circumstances it seemed wiser to avoid the risk of discouraging and thus spoiling the animal for use in the experiment. It should be stated, also, that it proved impossible to adhere to the period of thirty seconds as punishment in this series. For the majority of the wrong choices confinement of not more than ten seconds was used.

For the second series, given on May 4, the conditions were unfavorable in that it was dark and rainy, and the noise of the rain on the roof frightened Sobke. He refused to work after the fourth trial, and the series had to be completed on the following day. The total time required for this series was seventy-eight minutes.

The work on May 6 was distinctly better, and the animal's behavior indicated, in a number of trials, definite recognition of the right door. He might, for example, make a number of incorrect choices, then pause for a few seconds to look steadily at the doors, and having apparently found some cue, run directly to the right box. No aid from the experimenter was needed in this series.

On the following day improvement continued and the animal's method of choosing became definite and fairly precise. He was deliberate, quiet, and extremely business-like. The time for the series was thirty-one minutes.

The period of punishment was increased on May 12 to thirty seconds. Previously, for the greater number of the trials, it had been ten to fifteen seconds. This increase apparently did not disturb the monkey, for he continued to work perfectly throughout the series, although making many mistakes in spite of deliberate choices and the refusal of certain boxes in each trial.

An interesting and significant incident occurred on May 13 when at the conclusion of trial 5, Doctor Hamilton came into the experiment room for a few minutes. Sobke immediately stopped working, and he could not be induced to make any choices until Doctor Hamilton had left the room. This well indicates his sensitiveness to his surroundings, and his inclination to timidity or nervousness even in the presence of conditions not in themselves startling.

Work was continued thus steadily until May 28 when, because of the failure of the animal to improve, it seemed wise to increase the period of confinement as punishment to sixty seconds. In the meantime, it had sometimes been evident that Sobke was near to the solution of his problem. He would often make correct choices in three or four trials in succession and then apparently lose his cue and fail utterly for a number of trials.

After June 1, in order to hasten the solution of the problem, two series per day were given. In some instances the second series was given almost immediately after the first, while in others an interval of an hour or more intervened. It was further found desirable to give Sobke all of his food in the apparatus. When the rewards obtained in the several trials did not satisfy his hunger, additional food was presented, on the completion of the series of experiments, in one or more of the food cups. On days marked by unwillingness or refusal to work, very little food was given. Thus, the eagerness of the monkey to locate the right box was increased and, as a matter of observation, his deliberateness and care in choice increased correspondingly. Sixty seconds punishment was found satisfactory, and it was therefore continued throughout the work on this problem.

It was evident, on June 9, from the behavior of the monkey as well as from the score, that the perfect solution of the problem was near at hand. This fact the experimenter recorded in his daily notes, and sure enough, on the following day Sobke chose correctly throughout the series of ten trials. The time for this series was only ten minutes. The choices were made deliberately and readily.

An analysis of the data of table 5 reveals five methods or reactive tendencies which appeared more or less definitely in the following order: (a) Choice of first box at the left, because of experience in problem 1. This tendency was very quickly suppressed by the requirements in connection with problem 2. Indeed one of the most significant differences which I have discovered between the behavior of the primates and that of other mammals is the time required for the suppression of such an acquired tendency. The monkey seems to learn almost immediately that it is not worth while to persist in a tendency which although previously profitable no longer yields satisfaction, whereas in the crow, pig, rat, and ring dove, the unprofitable mode of response tends to persist during a relatively large number of trials. (b) The tendency to choose, first, a box near the left end of the group, to go from that to the box at the extreme right end of the group, thence to the one next in order, which was, of course, the right box. This tendency appears fairly clearly from May 7th on. (c) The box at the extreme right was first chosen and then the one next to it. For example, in setting 2, box 4 would be chosen first, then box 3. Or, if this did not occur, the method previously described under (b) was likely to be employed, as for example, in setting 8, where such choices as 7.6.5.1.8 appear. (d) In certain series there appeared a marked preference for a particular box, usually box 3 (see results for May 24). This was doubtless due in a measure, if not wholly, to the fact that box 3 was the right box twice in each series of ten settings. But it should be added that the same is true of box 7, for which no preference was manifested at any time. (e) Direct choice of the right box.

The five reactive methods or tendencies enumerated above roughly appeared in the order named, but there were certain irregularities and the order as well as the time of appearance varied somewhat from setting to setting. In general, method c was the most frequently used prior to the development of method e, the direct choice of the right box.

TABLE 5

Results for Sobke, _P. rhesus_, in Problem 2

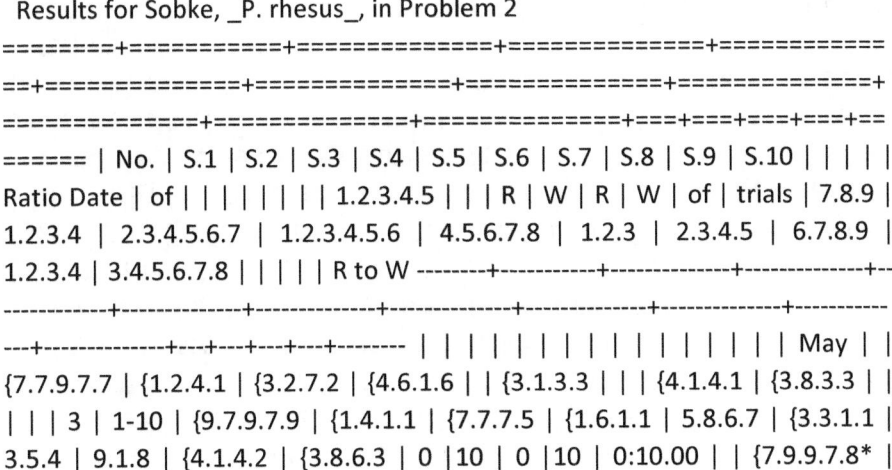

{4.3* | {2.7.6* | {1.3.5* | | {1.1.2* | | | {1.4.3* | {8.3.7* | | | | | | | | | | | | | | | | | | | | | | | | | | {4.7.2.7 | | (8.8.4.8 | | {5.5.5.5 | {9.5.5.9 | | | | | | | 4 and 5 | 11-20 | (7.9.7.9.9 | 1.4.1.3 | {2.7.3.2 | 1.6.5 | {8.8.4.8 | {3.3.1.1.3 | {5.5.5.3 | {7.1.6.6 | {4.4.4.4 | {8.8.6.8 | 0 |10 | 0 |10 | 0:10.00 | | {9.7.7.9.8 | | {7.2.6* | | | {4.8.7* | {3.3.1.2 | {5.5.4* | {6.5.8* | {4.4.4.3 | {8.7 | | | | | | | | | | | | | | | | | | | | | | | 6 | 21-30 | 9.7.9.7.8 | 4.3 | 7.2.3.7.6 | {6.3.1.6 | 6.8.7 | 3.1.2 | {5.3.5.3 | {6.9.4.6.1 | {4.1.2.4 | {3.4.5.3 | 0 |10 | 0 |10 | 0:10.00 | | | | | {6.2.6.5 | | | {2.4 | {7.9.7.8 | {4.2.3 | {8.6.7 | | | | | | | | | | | | | | | | | | | | | 7 | 31-40 | 7.9.7.8 | 1.4.3 | 2.7.6 | 3.5 | 4.8.7 | 3.1.2 | 3.2.3.5.4 | 8 | 4.3 | 5.8.3.4.8.7 | 1 | 9 | 1 | 9 | 1: 9.00 8 | 41-50 | 7.9.7.8 | (4.2.4.1.4 | 6 | 5 | 6.5.4.8.7 | {3.1.3.1 | 5.3.5.4 | {7.4.2.1 | {4.2.4.1 | 5.3.8.7 | 2 | 8 | 2 | 8 | 1: 4.00 | | | {4.2.2.4.3 | | | | {3.1.3.2 | | {9.8 | {4.3 | | | | | | | | | | | | | | | {5.3.5.3 | | | | | | | | | 10 | 51-60 | 7.7.8 | 3 | 7.3.2.7.6 | 6.4.6.5 | (6.4.8.5 | 3.1.3.2 | {2.5.3.5 | 5.9.8 | 4.2.4.3 | {6.5.4 | 1 | 9 | 1 | 9 | 1: 9.00 | | | | | {4.8.7 | | {2.5.4 | | | {3.8.7 | | | | | | | | | | | | | | | | | | | | | | | 11 | 61-70 | 7.9.7.8 | 1.4.3 | 4.3.2.7.6 | 6.5 | {5.4.8.6 | 3.2 | 5.4 | 6.5.2.3.1.8 | 3 | 6.5.3.8.7 | 1 | 9 | 1 | 9 | 1: 9.00 | | | | | | {5.4.8.7 | | | | | | | | | | | | | | | | | | | | | | 12 | 71-80 | 7.9.8 | 3 | 7.3.2.7.6 | 6.5 | 5.4.8.7 | 1.2 | 5.4 | 8 | 4.3 | 5.4.3.8.7 | 2 | 8 | 2 | 8 | 1: 4.00 13 | 81-90 | 7.8 | 4.3 | 3.7.6 | 6.5 | 8.7 | 1.3.2 | 2.5.2.5.4 | 7.6.5.9.8 | 3 | 8.7 | 1 | 9 | 1 | 9 | 1: 9.00 | | | | | | | | | | | | | | | | | | | | | | | | {1.2.1.2.2 | | | | | 14 | 91-100 | 7.8 | 3 | 6 | 5 | 6.5.4.7 | 1.3.2 | 5.4 | 9.8 | {1.4.2.2.1 | 7 | 4 | 6 | 4 | 6 | 1: 1.50 | | | | | | | | | {4.1.4.3 | | | | | | | | | | | | | | | | | | | | | | | 15 | 101-110 | 7.8 | 3 | 7.6 | 5 | 5.4.5.7 | 3.2 | 5.4 | 8 | 2.1.3 | 8.8.7 | 3 | 7 | 3 | 7 | 1: 2.33 | | | | | | | | | | | | | | | 17 | 111-120 | 7.8 | 4.3 | 7.6 | {2.1.2.1.2 | 8.7 | 3.2 | 5.4 | 7.6.5.1.8 | 3 | 8.7 | 1 | 9 | 1 | 9 | 1: 9.00 | | | | | {1.3.2.6.5 | | | | | | | | | | | | | | | | | | | | | | | | | 18 | 121-130 | 8 | 4.3 | 7.6 | 6.5 | 7 | 3.2 | 5.4 | 7.6.5.2.8 | 3 | 8.7 | 3 | 7 | 3 | 7 | 1: 2.33 | | | | | | | | | | | | | | | | | | | | | | | | | | {5.3.2.5.3 | | | | | | | | | | | | | | | | | {2.5.2.5.3 | | | | | | | 19 | 131-140 | 8 | 3 | 3.2.6 | 5 | 7 | 3.2 | {5.5.2.5.5 | 7.6.9.8 | 3 | 7 | 6 | 4 | 6 | 4 | 1: 0.67 | | | | | | | | | {3.2.5.5.3 | | | | | | | | | | | | | | | | | | {2.5.5.4 | | | | | | | | | | | | | | | | | | | | | | | | 20 | 141-150 | 7.8 | 3 | 3.2.7.6 | 6.5 | 5.4.7 | 3.2 | 3.2.5.4 | 6.5.4.3.8 | 3 | 7 | 3 | 7 | 3 | 7 | 1: 2.33 21 | 151-160 | 7.8 | 3 | 5.7.5.3.7.6 | 5 | 7 | 3.2 | 5.3.5.2.5.4 | 7.6.5.9.8 | 3 | 6.4.3.8.7 | 4 | 6 | 4 | 6 | 1: 1.50 | | | | | | | | | | | | | | | | 22 | 161-170 | 7.8 | 3 | 7.6 | 3.2.6.5 | 7 | 3.2 | {5.2.5.3 | 7.6.5.8 | 3 | 7 | 4 | 6 | 4 | 6 | 1: 1.50 | | | | | | | {2.5.4 | | | | | | | | | | | | | | | | | {3.1.3 | | | | | | | | | | 24 | 171-180 | 8 | 3 | {3.2.7.5 | {3.6.4 | 8.7 | {1.3.3 | {3.5.4 | 8 | {4.4.2.1 | 7 | 4 | 6 | 4 | 6 | 1: 1.50 | | | |

{7.3.6 | {3.6.5 | | {3.2 | | | {4.3 | | | | | | | | | | | | | | | | | | | | | | | 25 | 181-190 | 7.9.7.9.8 | 3 | 3.7.6 | 5 | 8.7 | 3.2 | 5.4 | 4.3.8 | 4.3 | 8.7 | 2 | 8 | 2 | 8 | 1: 4.00 26 | 191-200 | 8 | 3 | 6 | 5 | 8.7 | 3.2 | 5.4 | 8 | 4.3 | 8.7 | 5 | 5 | 5 | 5 | 1: 1.00 27 | 201-210 | 7.9.8 | 3 | 3.2.7.6 | 5 | 8.7 | 3.2 | 5.4 | 8 | 3 | 8.7 | 4 | 6 | 4 | 6 | 1: 1.50 28 | 211-220 | 8 | 3 | 3.7.6 | 4.3.2.6.5 | 8.7 | 3.2 | 5.4 | 7.6.5.9.8 | 3 | 7 | 4 | 6 | 4 | 6 | 1: 1.50 29 | 221-230 | 7.8 | 3 | 5.4.3.7.6 | 6.5 | 7 | 2 | 3.2.5.4 | 7.6.5.8 | 3 | 3.8.7 | 4 | 6 | 4 | 6 | 1: 1.50 31 | 231-240 | 7.7.8 | 3 | 3.7.6 | 6.5 | 8.7 | 3.2 | 5.4 | 9.8 | 3 | 3.7 | 2 | 8 | 2 | 8 | 1: 4.00 June | | | | | | | | | | | | | | | 1 | 241-250 | 8 | 3 | 5.4.7.6 | 3.2.6.5 | 8.7 | 3.2 | 5.4 | 8 | 3 | 8.7 | 4 | 6 | | | | | | | | | | | | | | | | " | 251-260 | 7.9.8 | 3 | 3.7.6 | 5 | 6.5.4.8.7 | 3.2 | 4 | 5.4.3.8 | {4.2.1.4 | 7 | 4 | 6 | 8 |12 | 1: 1.50 | | | | | | | | | | {2.4.4.3 | | | | | | | | | | | | | | | | | | | 2 | 261-270 | 7.8 | 3 | 6 | 3.2.6.5 | 7 | 3.2 | 3.4 | 8 | 3 | 6.5.3.8.7 | 5 | 5 | | | " | 271-280 | 7.8 | 3 | 3.7.6 | 3.6.5 | 7 | 3.2 | 5.4 | 7.6.5.4.3.8 | 3 | 7 | 4 | 6 | 9 |11 | 1: 1.22 3 | 281-290 | 7.8 | 3 | 7.6 | 3.6.5 | 8.7 | 3.2 | 4 | 7.8 | 3 | 7 | 4 | 6 | | | " | 291-300 | 9.8 | 3 | 3.6 | 4.3.6.5 | 8.7 | 2 | 3.5.4 | 7.6.5.9.8 | 3 | 8.7 | 3 | 7 | 7 |13 | 1: 1.86 4 | 301-310 | 8 | 3 | 7.6 | 3.4.3.6.5 | 8.7 | 2 | 3.2.5.4 | 7.6.5.8 | 3 | 3.8.7 | 4 | 6 | | | " | 311-320 | 8 | 3 | 5.4.7.6 | 3.2.6.5 | 7 | 2 | 3.2.5.4 | 8 | 3 | 7 | 7 | 3 |11 | 9 | 1: 0.82 5 | 321-330 | 8 | 3 | 6 | 4.6.5 | 7 | 2 | 3.5.4 | 8 | 4.3 | 7 | 7 | 3 | | | " | 331-340 | 8 | 3 | 7.4.7.6 | 3.2.4.6.5 | 7 | 3.2 | 3.5.4 | 8 | 3 | 7 | 6 | 4 |13 | 7 | 1: 0.54 7 | 341-350 | 8 | 3 | 7.6 | 5 | 8.7 | 1.3.1.2 | 3.5.4 | 8 | 3 | 7 | 6 | 4 | | | " | 351-360 | 8 | 3 | 7.6 | 5 | 8.7 | 2 | 3.5.4 | 8 | 3 | 7 | 7 | 3 |13 | 7 | 1: 0.54 8 | 361-370 | 7.8 | 3 | 4.7.6 | 3.5 | 8.7 | 2 | 3.4 | 9.8 | 3 | 7 | 4 | 6 | | | | 371-380 | 8 | 3 | 7.6 | 3.4.4.3.5 | 8.7 | 3.2 | 3.4 | 8 | 3 | 7 | 5 | 5 | 9 |11 | 1: 1.22 9 | 381-390 | 8 | 3 | 6 | 4.2.1.5 | 7 | 2 | 3.4 | 8 | 3 | 8.7 | 7 | 3 | | | " | 391-400 | 8 | 3 | 6 | 5 | 7 | 2 | 3.4 | 8 | 3 | 7 | 9 | 1 |16 | 4 | 1: 0.25 10 | 401-410 | 8 | 3 | 6 | 5 | 7 | 2 | 4 | 8 | 3 | 7 |10 | 0 |10 | 0 | 1: 0.00 --------+-----------+--------------+--------------+--------------+--------------+--------------+--------------+--------------+--------------+---+---+---+---+-------- | | | | 1.2.3.4.5 | | | | | | 1.2.3.4 | 2.3.4.5 | | | | | | | 5.6.7.8 | 2.3.4.5.6 | 6.7.8.9 | 5.6.7 | 1.2.3.4 | 4.5.6 | 2.3.4.5 | 1.2.3 | 5.6.7 | 6.7.8.9 | | | | | | +--------------+--------------+--------------+--------------+--------------+--------------+--------------+--------------+--------------+---+---+---+---+---- ---- 11 | 1-10 | 6.7 | 3.5 | 8 | 6 | 3 | 5 | 3.4 | 3.2 | 7.7.2.6 | 8 | 5 | 5 | 5 | 5 | 1: 1.00 12 | 11-20 | 7 | 3.6.6.2.5 | 8 | 6 | 3 | 4.5 | 4 | 2 | 7.6 | 8 | 7 | 3 | 7 | 3 | | 1: 0.43
=======+===========+==============+==============+============== ==+==============+==============+==============+==============+

```
==============+==============+==============+===+===+===+===+==
======
```

[Footnote *: Aided by experimenter.]

Examination of table 5 indicates that some of the settings proved very easy for Sobke; others, extremely difficult. Consequently, the number of methods which were tried and rejected for a given setting varies from two to five. Setting 2 proved a fairly simple one, and after the inhibition of the tendency to choose the first box at the left, the only definite tendency to appear was that of choosing the first box at the right, and then the one next to it. After one hundred and thirty trials, this method suddenly gave place to direct choice of the right box, and during the following twenty-eight series, no error occurred for this setting. Setting 4, on the contrary, proved extremely difficult, and a variety of methods is more or less definitely indicated by the records.

It is needless to lengthen the description by analyzing the data for each setting, since the reader by carefully scanning the columns of data in table 5 may observe for himself the various tendencies and their mutual relations.

Sobke's curve of learning (figure 19) in problem 2, is extremely irregular, as was that of Skirrl. Similar irregularities appear in the daily ratios of right to wrong first choices presented in the last column of table 5. Most of these irregularities were due, I have discovered, to unfavorable external conditions. Thus, dark rainy days and disturbing noises outside the laboratory were obviously conditions of poor work.

On the day following the final and correct series for problem 2, a control series was given. In this Sobke seemed greatly surprised by the new situations which presented themselves. Repeatedly he exhibited impulses to enter the box which would have been the correct one in the regular series of settings. He frequently inhibited such impulses and chose correctly, but at other times he reacted quickly and made mistakes. It was evident from his behavior that he was not guided by anything like a definite idea of the relation of the right box to the other members of the group.

In a second control series given on the following day, June 12, confusion appeared, but less markedly. For the first setting, a correct choice was made with deliberation. For the second setting, box 3 was immediately chosen, as should have been the case in the regular series of settings. Sobke seemed confused when he emerged from this box and had difficulty in locating the right one. Then followed direct correct choices for settings 3, 4, and 5. For setting 6, there is recorded a deliberately made wrong choice, and so on throughout the series, the choices being characterized by deliberateness and definite search for the right box. Uncertainty was plainly indicated, and in this the behavior of the animal differed markedly from that in the concluding series of the regular experiment.

It seems safe to conclude from the results of these control series that Sobke has no free idea of the relation of secondness from the right and is chiefly dependent upon memory of the particular settings for cues which lead to correct choice.

Problem 3. Alternately First at Left and First at Right

For four successive days after the last control series in connection with problem 2, Sobke was merely fed in the apparatus according to previous description (p. 43). He exhibited a wonderfully keen appetite and was well fed during this interval between problems.

The method of experimentation chosen for problem 3 in the light of previous experience was that of confining the monkey for a short time, ten to fifteen seconds, in the wrong box, in each of the first ten mistakes for a given trial, and of then aiding him to find the right box by the slight and momentary raising of the exit door. Aid proved necessary in a few of the trials during the first four days. Then he worked independently. As work progressed it was found possible and also desirable to increase the period of confinement, and in the end, sixty seconds proved satisfactory. It was also thought desirable to increase the number of trials per day from a single series during the early days to two or even three series from June 29 on. Often three series could be given in succession without difficulty. During the early trials on this problem Sobke worked remarkably well, but later his willingness diminished, evidently because of his failure readily to solve the problem, and it became extremely difficult to coax him into the apparatus. On days when he entered only reluctantly and as it seemed against his will, he was likely to be nervous, erratic, and often slow in making his choices, but above all he tended to waste time by not returning to the starting point, preferring rather to loiter in the alleyways or run back and forth.

TABLE 6

Results for Sobke, _P. rhesus_, in Problem 3

```
========+===========+==============+==============+============
==+=============+==============+==============+==============+
=============+==============+==============+===+===+===+===+==
====== | No. | S.1 | S.2 | S.3 | S.4 | S.5 | S.6 | S.7 | S.8 | S.9 | S.10 | | | |
Ratio Date | of | | | | | | | | | 3.4.5.6 | 3.4.5.6 | R | W | R | W | of | trials |
5.6.7 | 5.6.7 | 1.2.3.4.5.6 | 1.2.3.4.5.6 | 4.5.6.7.8 | 4.5.6.7.8 | 2.3.4.5 |
2.3.4.5 | 7.8.9 | 7.8.9 | | | | | R to W --------+-----------+--------------+--------------
+--------------+--------------+--------------+--------------+--------------+--------
------+--------------+---+---+---+---+-------- | | | | | | | | | | | | | | | June | | | |
| {3.1.2.1 | {7.8.7.8 | | | | {9.8.7.9 | | | | | | 17 | 1--10 | {6.6.7.6 | 5.7 |
{4.4.3.5.4 | {4.4.2.1 | {8.8.7.8 | 8 | 2 | 3.2.5 | {8.7.6.9 | 3.8.9 | 2 | 8 | 2 | 8 |
1:4.00 | | {6.5 | | {5.4.2.1 | {5.4.6* | {7.7.4* | | | | {8.7.3* | | | | | | | | | | | |
| | | | | | | | | | | | | | {3.4.2.5 | | {8.7.7.5 | | | | | | | | | 18 | 11--20 |
6.6.6.5 | 5.7 | {4.5.6.4 | 3.1.4.3.6 | {8.8.7.8 | 8 | 3.2 | (3.4.2.4 | 9.8.8.7.3 |
8.6.9 | 1 | 9 | 1 | 9 | 1:9.00 | | | | {2.3.1* | | {7.8.4* | | | {4.5 | | | | | | | | |
| | | | | | | | | | | | | | | | | {3.6.5.4 | | {8.8.7.7 | | | | | | | | | 19 | 21--
30 | 6.5 | 7 | {2.5.2.6 | 3.6 | {7.8.8.8 | 8 | 4.3.2 | {3.2.4.2 | {9.8.7.9.7 | 9 | 3 |
7 | 3 | 7 | 1:2.33 | | | | {3.5.1* | | {8.4 | | | {2.5 | {6.5.4.3 | | | | | | | | | |
| | | | | | | | | | | | | | | | | (8.8.8.8 | | | | {9.9.8.9 | | | | | | 21 | 31--40 |
6.5 | 5.6.5.7 | {3.5.4.6 | 2.5.3.6 | {7.8.3.8 | 8 | 4.2 | 2.5 | {9.5.9.9 | 9 | 2 | 8 |
2 | 8 | 1:4.00 | | | | {4.3.5.1 | | {8.7.4* | | | | {9.9.3* | | | | | | | | | | | | | |
| | | | | | | | 22 | 41--50 | 7.6.5 | 6.5.5.7 | 1 | 2.1.3.6 | {8.8.8.8 | 8 | 2 |
2.3.2.2.5 | {9.8.9.8.7 | 9 | 4 | 6 | 4 | 6 | 1:1.50 | | | | | | {7.8.4 | | | | {6.5.9.3
| | | | | | | | | | | | | | | | | | | | | | | | 23 | 51--60 | 5 | 6.5.7 | 1 | {1.2.1.1 |
8.5.4 | 8 | 5.4.2 | 2.4.2.5 | {9.8.4.9 | 9 | 4 | 6 | 4 | 6 | 1:1.50 | | | | | {3.2.6 |
| | | | {7.3 | | | | | | | | | | | | | | | | | | | | | | | 24 | 61--70 | 7.6.5 | 7 | 2.3.1
| 2.1.5.4.2.6 | 8.7.8.4 | 8 | 4.5.4.3.2 | 2.2.4.5 | 9.7.6.8.3 | 9 | 3 | 7 | 3 | 7 |
1:2.33 25 | 71--80 | 6.5 | 5.7 | 3.1 | 6 | 8.5.4 | 8 | 2 | 3.2.5 | 9.8.7.3 | 3.9 | 3
| 7 | 3 | 7 | 1:2.33 | | | | | | | | | | | | | | | | | 26 | 81--90 | 7.7.6.5 | 6.5.7 |
3.1 | {1.4.1.1 | 8.4 | 8 | 5.4.2 | 2.5 | 9.8.9.6.3 | 9 | 2 | 8 | 2 | 8 | 1:4.00 | | |
| | {5.1.6 | | | | | | | | | | | | | | | | | | | | | | | | | | | | | | 28 | 91--100 | 7.6.5 |
```

7 | 1 | 1.2.4.6 | 8.4 | 8 | 2 | 3.2.2.2.5 | 9.8.8.7.9.3 | 9 | 5 | 5 | 5 | 5 | 1:1.00 |
| | | | | | | | | | | | | 29 | 101--110 | 7.6.5 | 5.7 | 1 | {1.1.5.3 | 8.4 | 8 | 2
| 4.2.5 | {9.9.8.9.4 | 9 | 4 | 6 | | | | | | | | {2.1.6 | | | | | {9.7.6.3 | | | | | | |
| | | | | | | | | | | | | | | | " | 111--120 | 6.5 | 6.5.7 | 1 | 2.1.1.3.6 | 8.5.4 | 8 |
2 | 3.2.5 | 9.8.7.6.4.3 | 9 | 4 | 6 | 8 |12 | 1:1.50 30 | 121--130 | 5 |
6.5.6.5.5.7 | 1 | 3.1.6 | 8.7.6.8.5.4 | 8 | 2 | 4.2.4.2.5 | 9.3 | 9 | 5 | 5 | | | " |
131--140 | 5 | 7 | 1 | 2.3.6 | 8.5.6.4 | 8 | 2 | 5 | 9.8.3 | 3.9 | 6 | 4 |11 | 9 |
1:0.82 July | | | | | | | | | | | | | | | | | | | 1 | 141--150 | 5 | 7 | 1 | 1.6 | 8.7.4 | 8
| 2 | 3.2.5 | 9.8.6.9.3 | 9 | 6 | 4 | | | " | 151--160 | 5 | 7 | 1 | 2.5.3.6 | 8.4 | 8
| 2 | 2.5 | 9.3 | 8.8.7.5.4.9 | 5 | 5 |11 | 9 | 1:0.82 2 | 161--170 | 6.5 | 7 | 1 |
2.6 | 8.4 | 8 | 2 | 3.5 | 9.3 | 9 | 5 | 5 | 5 | 5 | 1:1.00 3 | 171--180 | 6.5 | 7 | 1
| 1.5.6 | 8.4 | 8 | 2 | 3.5 | 9.3 | 9 | 5 | 5 | | | " | 181--190 | 5 | 7 | 1 | 4.6 |
8.6.4 | 8 | 2 | 5 | 9.8.4.8.5.3 | 9 | 7 | 3 |12 | 8 | 1:0.67 5 | 191--200 | 6.5 |
5.7 | 1 | 6 | 8.4 | 8 | 2 | 5 | 9.5.3 | 9 | 6 | 4 | | | " | 201--210 | 5 | 7 | 6.1 |
2.6 | 8.6.4 | 8 | 5.3.5.4.3.2 | 4.3.5 | 9.7.3 | 9 | 4 | 6 |10 |10 | 1:1.00 | | | |
| | | | | | | | | | 6 | 211--220 | 5 | 5.7 | 1 | 2.6 | 8.6.4 | 8 | 2 | 4.3.5 | 9.3 |
{5.4.8.3 | 4 | 6 | | | | | | | | | | | | | | {8.5.4.9 | | | | | | | | | | | | | | | | |
| | | " | 221--230 | 5 | 7 | 1 | 5.3.6 | 8.6.8.4 | 8 | 4.2 | 3.2.5 | 9.3 | 9 | 5 | 5 |
9 |11 | 1:1.22 7 | 231--240 | 5 | 7 | 1 | 2.6 | 8.4 | 7.4.8 | 2 | 4.3.5 | 8.3 | 9 |
5 | 5 | | | " | 241--250 | 5 | 7 | 2.6.3.5.1 | 6 | 8.4 | 8 | 2 | 5 | 9.3 | 9 | 7 | 3 |
| | | | | | | | | | | | | | | | | | | | | | {2.6.2.6.4 | | | | | | | | | | | " | 251--
260 | 5 | 7 | {6.3.2.6 | 6 | 8.4 | 8 | 2 | 5 | 9.5.3 | 9 | 7 | 3 |19 |11 | 1:0.58 | |
| | {5.4.3.1 | | | | | | | | | | | | | | | | | | | | | | | | | | | | 8 | 261--270 | 7.5
| 7 | 1 | 6 | 8.4 | 8 | 5.2 | 5 | 8.3 | 8.5.4.9 | 5 | 5 | | | " | 271--280 | 5 | 7 |
2.6.4.6.1 | 6 | 8.5.8.4 | 7.4.8 | 5.5.3.5.2 | 5 | 9.3 | 9 | 5 | 5 |10 |10 | 1:1.00 |
| | | | | | | | | | | | | 9 | 281--290 | 5 | 5.7 | 1 | 6 | 8.4 | 8 | 2 | 2.5 | 8.3
| {7.3.8.6.8 | 5 | 5 | | | | | | | | | | | | | | | {4.7.3.9 | | | | | | | | | | | | | |
| | | | | " | 291--300 | 5 | 7 | 1 | 6 | 8.4 | 8 | 2 | 5 | 9.3 | 9 | 8 | 2 | | | " |
301--310 | 5 | 7 | 1 | 6 | 8.4 | 8 | 5.2 | 5 | 9.5.3 | 9 | 7 | 3 |20 |10 | 1:0.50
10 | 311--320 | 5 | 7 | 1 | 6 | 8.4 | 7.4.8 | 2 | 5 | 9.3 | 9 | 7 | 3 | | | " | 321--
330 | 5 | 7 | 1 | 5.2.6 | 8.7.4 | 8 | 2 | 5 | 9.3 | 9 | 7 | 3 |14 | 6 | 1:0.43 12 |
331--340 | 5 | 7 | 1 | 6 | 8.4 | 8 | 2 | 5 | 8.3 | 9 | 8 | 2 | | | " | 341--350 | 5 |
7 | 1 | 6 | 8.4 | 8 | 5.2 | 5 | 9.3 | 9 | 7 | 3 |15 | 5 | 1:0.33 13 | 351--360 | 5 |
5.7 | 1 | 6 | 8.4 | 8 | 5.2 | 5 | 3 | 9 | 7 | 3 | | | " | 361--370 | 5 | 7 | 6.1 | 6 |
4 | 8 | 5.2 | 3.2.5 | 9.3 | 9 | 6 | 4 |13 | 7 | 1:0.54 14 | 371--380 | 5 | 7 | 1 | 6
| 8.4 | 8 | 2 | 5 | 3 | 7.4.3.9 | 8 | 2 | | | " | 381--390 | 5 | 7 | 1 | 6 | 8.4 | 8 |
2 | 5 | 9.3 | 4.7.3.9 | 7 | 3 |15 | 5 | 1:0.33 15 | 391--400 | 5 | 5.5.7 | 1 | 3.6 |
8.4 | 8 | 3.2 | 5 | 8.3 | 9 | 5 | 5 | | | " | 401--410 | 5 | 7 | 1 | 6 | 8.4 | 8 | 2 |

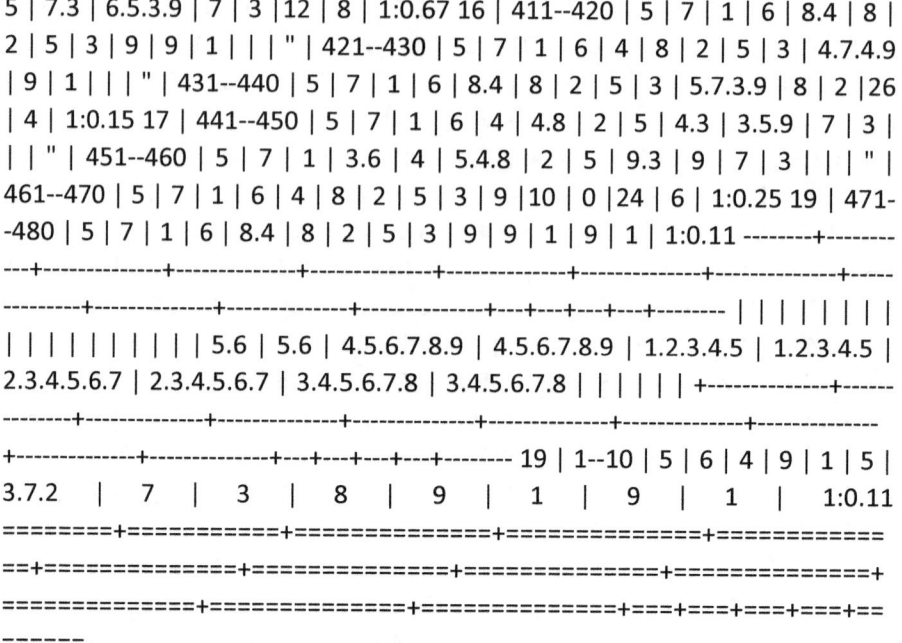

```
5 | 7.3 | 6.5.3.9 | 7 | 3 |12 | 8 | 1:0.67 16 | 411--420 | 5 | 7 | 1 | 6 | 8.4 | 8 |
2 | 5 | 3 | 9 | 9 | 1 | | | " | 421--430 | 5 | 7 | 1 | 6 | 4 | 8 | 2 | 5 | 3 | 4.7.4.9
| 9 | 1 | | | " | 431--440 | 5 | 7 | 1 | 6 | 8.4 | 8 | 2 | 5 | 3 | 5.7.3.9 | 8 | 2 |26
| 4 | 1:0.15 17 | 441--450 | 5 | 7 | 1 | 6 | 4 | 4.8 | 2 | 5 | 4.3 | 3.5.9 | 7 | 3 |
| | " | 451--460 | 5 | 7 | 1 | 3.6 | 4 | 5.4.8 | 2 | 5 | 9.3 | 9 | 7 | 3 | | | " |
461--470 | 5 | 7 | 1 | 6 | 4 | 8 | 2 | 5 | 3 | 9 |10 | 0 |24 | 6 | 1:0.25 19 | 471-
-480 | 5 | 7 | 1 | 6 | 8.4 | 8 | 2 | 5 | 3 | 9 | 9 | 1 | 9 | 1 | 1:0.11 --------+--------
---+--------------+--------------+--------------+--------------+--------------+--------------+-----
---------+--------------+--------------+--------------+---+---+---+---+-------- | | | | | | | |
| | | | | | | | | | 5.6 | 5.6 | 4.5.6.7.8.9 | 4.5.6.7.8.9 | 1.2.3.4.5 | 1.2.3.4.5 |
2.3.4.5.6.7 | 2.3.4.5.6.7 | 3.4.5.6.7.8 | 3.4.5.6.7.8 | | | | | | +--------------+------
--------+--------------+--------------+--------------+--------------+--------------+--------------
+--------------+--------------+---+---+---+---+-------- 19 | 1--10 | 5 | 6 | 4 | 9 | 1 | 5 |
3.7.2   |   7   |   3   |   8   |   9   |   1   |   9   |   1   |   1:0.11
========+===========+==============+==============+==============+===========
==+=============+==============+==============+==============+===============+
==============+==============+==============+===+===+===+===+==
======
```

[Footnote *: Aided by experimenter.]

The data of table 6 indicate for this problem only three pronounced reactive tendencies: (a) As the initial tendency, the choice of the second box from the right end. This proved surprisingly weak, in view of the animal's long training on problem 2, and it disappeared quickly. (b) Choice of the end boxes, and (c) direct choice of the right box.

For this, as for the other problems, extreme differences in method and in time and degree of success appear for the different settings. Thus, while settings 1, 2, 3, 6, 7, and 10 proved to be easy, settings 4, 5, 8, and 9 were evidently more difficult.

[Illustration: FIGURE 20.--Error curve of learning for the solution of problem 3 (alternately the first box at the left end and the first at the right end) by Sobke.]

From the first this problem promised to be much easier for Sobke than problem 2, and although the actual number of trials necessary for the solution is greater by sixty for problem 3 than for problem 2, comparison of the data of the tables justifies the statement that the third problem was both easier and more nearly adequately solved than the second. This is not surprising when the nature of the two problems is considered, for whereas problem 2 requires choice by perception of the relationship of secondness from the right end of the group, problem 3 requires, instead, the choice of the end member of the group each time, with the additional variation of alternation of ends. Now as it happens, the end member is easily selected by the monkey, and it appears further that alternation was relatively easy for Sobke to acquire. Consequently, the combination of end and alternation proved easier than the choice of the second from the right end of the group.

The above statements are supported by comparison of the curves of learning. The curve for problem 2, figure 19, is extremely irregular; that for problem 3, figure 20, much more regular. Similarly, the daily ratios of right to wrong choices as exhibited in tables 5 and 6 indicate smaller variations for the third problem than for the second.

Sobke made ten correct first choices in the third series for July 17, but he was working very uncertainly and it seemed rather a matter of good luck than good management that he succeeded in presenting this perfect series: For this reason and also because it did not seem feasible to have Sunday intervene between the final and perfect regular series and the control series, an additional regular series was given on July 19, in which, as the table indicates, a single mistake occurred, in trial 5. The monkey was working perfectly. The series of trials required only ten minutes, and it was evident that carelessness and eagerness to obtain food were chiefly responsible for the mistake.

The control series given on July 19 immediately after the series just described resulted similarly in one failure and nine successes. The choices were made easily and with certainty, and the only mistake, that of setting 7, was apparently due to carelessness.

This excellent showing for the control series wholly justifies the comparison of problems 2 and 3 as to difficultness, made above. Whereas in both problems 1 and 2 the control trials caused confusion, in the case of problem 3, they did not essentially alter the behavior of the animal. The fact seems to be that for this problem the particular setting is of relatively little importance; while turning alternately to the extreme left and the extreme right is of prime importance. That Sobke had the idea of alternation or of the end box, there seems no more reason for insisting than that he had the idea of secondness from the right end in problem 2. It is possible, even probable, that these ideas existed rather vaguely in his consciousness, but there is obviously no necessity for insisting that the solution of the problems depended upon them.

Problem 4. Middle

As the available time for the continuation of the experiment was limited, it was decided to proceed with work on problem 4 immediately upon the completion of problem 3, and on July 20, the problem of the middle door was presented to Sobke. Since it was anticipated that this sudden change would confuse and discourage him greatly, the only form of punishment administered was the momentary lowering of the entrance door of the wrong box. As in the previous problem, he was aided after ten successive wrong choices. As might have been anticipated, he persistently entered the end boxes of the groups, and this in some instances probably would have been kept up for many minutes had not the experimenter lured him into the right box by slightly raising the exit door. In the first series, he had to be aided in five of the ten trials. The total time for the series was forty-five minutes, the total number of choices, eighty-eight. In the second series, he was aided in four of the trials. The total time required was seventy-two minutes, and the total number of choices was seventy-six.

Throughout the first series, Sobke worked hard, but with evidently increasing dissatisfaction. He clung persistently to his acquired tendency to choose the end boxes, and after each trial he returned less willingly to the starting point. Up to this time his attitude toward the experimenter had been perfectly friendly, if not wholly trustful. But when on July 21 he was brought into the apparatus for the second series, he exhibited a wholly new form of behavior, for instead of attending diligently to the open doors and devoting his energies to trying to find the right box, he instead, after gazing at them for a few seconds, turned toward the experimenter and jumped for him savagely, throwing himself against the wire netting with great force. This was repeated a number of times during the first two or three trials, and it occurred less frequently later in the series. Since nothing unusual had happened outside of the experiment room, the suggested explanation of this sudden change in attitude and behavior is that the monkey resented and blamed on the experimenter the difficulty which he was having in obtaining food.

From this time on until the end of my work, Sobke was always savage and both in and out of the apparatus he was constantly on the watch for an opportunity to spring upon me. Previously, it had been possible for me to coax him into the apparatus by offering him food and to return him to his cage by walking after him. But on and after the twenty-first of July, it was impossible for me to approach him without extreme risk of being bitten.

Doctor Hamilton when told of this behavior, reported that several times monkeys have shown resentment toward him when they were having trouble in the experiment. I therefore feel fairly confident that I have not misinterpreted Sobke's behavior. When on July 22 I gave Sobke an opportunity to enter the apparatus, he refused, and it was impossible to lure him in with food. Two hours later, having waited meantime for his breakfast, he entered readily and worked steadily and persistently through his third series of trials, but in no one of these trials did he choose correctly. Neither on this day nor the following did he exhibit resentment while at work. He apparently had regained his affective poise and was able to attend as formerly to the task of locating his rewards.

During these first three series, although the ratio of right to wrong choices stood 0 to 10, there occurred a marked reduction in the number of trials in which aid was necessary as well as in the total number of choices, and on July 23 correct reactions began to appear. Improvement during the next hundred trials was steady and fairly rapid, and on July 31, a record of seven right to three wrong trials was obtained. This was surprising to the experimenter, as well as gratifying, since he was eager to have the animal complete this problem before work should have to be discontinued.

Everything went smoothly until August 2, when my assistant, who had been left in charge of the experimental work for a week, attempted to increase the number of trials per day to two series. Sobke apparently was not quite ready for this increase in the amount of his day's labor and refused to work at the end of the first series. In this series he did less well than on the previous day. The following day, August 3, unfortunately and contrary to the wishes of the experimenter, the laboratory was painted and there was necessarily considerable disturbance because of the presence of the workmen, and in addition, the pervasive odor of fresh paint. Sobke chose still less successfully on this date, and on August 4, he refused to work after the eighth trial. It is true that during these bad days the total number of choices steadily diminished while the successes, also, diminished, or at best, failed to increase. When on August 9, I returned to the laboratory to take charge, I found that Sobke was no longer trying to solve the problem as when I had gone away. His attitude had changed in that he had become indifferent, careless, and obviously discouraged with his task.

TABLE 7

Results for Sobke, _P. rhesus_, in Problem 4

```
========+===========+==============+==============+============
==+=============+==============+==============+==============+
============+==============+==============+===+===+===+===+==
====== | No. | S.1 | S.2 | S.3 | S.4 | S.5 | S.6 | S.7 | S.8 | S.9 | S.10 | | | | |
Ratio Date | of | | | 1.2.3.4 | | | 1.2.3.4.5 | | | 3.4.5.6 | | R | W | R | W | of |
trials | 2.3.4 | 5.6.7.8.9 | 5.6.7 | 7.8.9 | 4.5.6.7.8 | 6.7.8.9 | 1.2.3 | 2.3.4.5.6 |
7.8.9 | 6.7.8 | | | | | R to W --------+-----------+--------------+--------------+-------------
--+--------------+--------------+--------------+--------------+--------------+------
--------+---+---+---+---+-------- | | | | | | | | | | | | | | | | | | July | | | {5.9.5.5 |
{1.7.1.3 | {9.7.9.7 | | {1.9.3.1 | | | {6.3.2.6 | | | | | | | | 20 | 1- 10 | 2.4.2.4.3 |
{5.9.5.6 | {1.7.1.7 | {7.9.7.9 | {8.4.8.4 | {9.2.9.3 | 3.1.3.2 | {3.6.3.2 | {9.3.4.3 |
{8.6.6.8.6 | 0 |10 | 0 |10 | 0:10.00 | | | {5.8.7* | {1.7.4* | {7.7.8 | {4.6 |
{9.1.5* | | {6.3.4* | {3.9.3.6 | {8.6.8.6.7 | | | | | | | | | | | | | | | | | | | | | | | |
| | | | {7.9.7.7 | {4.8.5.5 | {1.4.3.2 | | | {3.9.3.8 | | | | | | 21 | 11- 20 | 2.3 |
{5.6.5.5 | {1.2.7.1 | {7.9.7.7 | {8.4.8.4 | {8.9.1.9 | 3.1.2 | {6.2.3.2.6 | {4.3.3.5 |
6.7 | 0 |10 | 0 |10 | 0:10.00 | | | {9.5.5.7 | {7.3.7.4 | {9.7.8* | {5.4.6* |
{1.9.5* | | {3.5.2.4 | {3.4.6* | | | | | | | | | | | | | | | | | | | | | | | | | | | | |
| | | {3.7.4.8 | | | | | | 22 | 21- 30 | 2.3 | 5.6.5.6.7 | 1.7.4 | {7.9.7 | 4.7.4.6 |
{1.4.6.3 | 3.1.2 | 5.2.6.4 | {4.3.5.8 | 6.6.7 | 0 |10 | 0 |10 | 0:10.00 | | | | |
{7.7.8 | | {2.7.5 | | | {3.7.6* | | | | | | | | | | | | | | | | | | | | | | | | | | | | | |
| | | {5.4.7.4 | | | | | | 23 | 31- 40 | 2.4.3 | 5.6.8.7 | 1.7.4 | {7.7.7.7 | 4.7.6 |
2.7.5 | 3.1.2 | 6.4 | {7.5.4.3 | 7 | 1 | 9 | 1 | 9 | 1: 9.00 | | | | | {7.7.8 | | | | |
{7.3.6 | | | | | | | | | | | | | | | | | | | | | | | | 24 | 41- 50 | 3 | 5.6.7 | 1.6.4 |
```

7.7.7.8 | 4.7.6 | {2.7.6.4 | 3.2 | 6.5.4 | 5.3.8.6.6 | 7 | 1 | 9 | 1 | 9 | 1: 9.00 | |
| | | | | {2.7.3.8 | | | | | | | | | | | | | | | | | | | | | {9.6.5* | | | | | | | | | | | | | | | |
| | | | | | | | | 26 | 51- 60 | 4.3 | 6.5.7 | 2.7.4 | 7.8 | 6 | {6.4.2.7 | 3.2 |
{6.5.3.5.2 | 7.6 | 7 | 2 | 8 | 2 | 8 | 1: 4.00 | | | | | | | {4.8.6.5 | | {6.5.6.4 | |
| | | | | | | | | | | | | | | | | | | | | | | | | | | | | {4.7.3.7 | | | | | | | | | 27 |
61- 70 | 3 | 6.5.7 | 2.5.4 | 7.7.8 | 5.7.6 | {4.6.1.4 | 2 | 6.5.4 | 5.7.6 | 6.7 | 2 |
8 | 2 | 8 | 1: 4.00 | | | | | | | {7.3.5 | | | | | | | | | | | | | | | | | | | | | | |
28 | 71- 80 | 3 | 7 | 6.5.4 | 7.8 | 5.4.7.6 | 2.7.6.5 | 2 | 5.5.4 | 7.6 | 7 | 4 | 6 |
4 | 6 | 1: 1.50 | | | | | | | | | | | | | | | | 29 | 81- 90 | 3 | 6.5.7 | 2.4 | 7.7.8 |
5.4.6 | {2.7.7.6 | 2 | 4 | 5.4.7.6 | 7 | 4 | 6 | 4 | 6 | 1: 1.50 | | | | | | | {2.7.6.5
| | | | | | | | | | | | | | | | | | | | | | | | | | | 30 | 91-100 | 3 | 7 | 2.6.5.4 | 7.8 |
6 | 5 | 2 | 5.4 | 5.4.6 | 7 | 6 | 4 | 6 | 4 | 1: 0.67 31 | 101-110 | 3 | 7 | 2.4 |
7.8 | 6 | 5 | 2 | 4 | 7.6 | 7 | 7 | 3 | 7 | 3 | 1: 0.43 August | | | | | | | | | | |
| | | | 2 | 111-120 | 3 | 7 | 6.5.7.6.4 | 7.8 | 6 | 4.2.7.6.5 | 2 | 6.5.4 | 7.6 | 7 |
5 | 5 | 5 | 5 | 1: 1.00 3 | 121-130 | 3 | 6.5.7 | 7.6.5.4 | 7.8 | 7.6 | 5 | 2 | 6.5.4
| 7.6 | 7 | 4 | 6 | 4 | 6 | 1: 1.50 | | | | | | | | | | | | | | | 4 | 131-140 | 3 | 7
| 6.5.7.6.4 | 7.8 | 6 | {2.7.6.4 | 2 | 6.4 | 3.5.4.6 | 7 | 5 | 5 | 5 | 5 | 1: 1.00 | |
| | | | | {8.7.6.5 | | | | | | | | | | | {2.4.4.4 | | | | | | | | | | | | | | | 5 | 141-
150 | {2.4.4.2 | 6.5.7 | {2.7.6.7 | 8 | 7.6 | {2.8.7.6 | 3.2 | 6.4 | 7.6 | 8.7 | 1 | 9
| 1 | 9 | 1: 9.00 | | {4.3* | | {5.4 | | | {8.6.5 | | | | | | | | | | | | | | | | | | | |
| | | | | 6 | 151-160 | 2.4.4.3 | 7 | 2.6.5.4 | 7.8 | 7.6 | 7.6.5 | 2 | 6.4 | 6 | 7 |
4 | 6 | | | | | | | | | | | | | | | | | | | | | | | | | | {4.4.2.4 | | | | | | | | | | | " |
161-170 | {4.2.4.2 | 7 | 6.5.4 | 7.8 | 7.6 | 7.6.5 | 3.2 | 5.4 | 7.6 | 7 | 2 | 8 | 6
| 14 | 1: 2.33 | | {4.4.3 | | | | | | | | | | | | | | | | | | | | | | | | | | | | | 7 |
171-180 | 4.3 | 7 | 6.5.4 | 8 | 7.6 | 5 | 2 | 6.5.4 | 7.6 | 7 | 5 | 5 | | | | | | | |
| | | | | | | | | | | " | 181-190 | {4.2.4.4 | 7 | 7.6.5.4 | 7.8 | 6 | 6.5 | 2 | 6.5.4
| 7.6 | 7 | 4 | 6 | 9 | 11 | 1: 1.22 | | {2.4.3 | | | | | | | | | | | | | | | | | | | |
| | | | | | | | | 9 | 191-200 | 3 | 7 | 5.4 | 8 | 8.7.6 | 6.5 | 2 | 6.5.4 | 7.6 | 8.7
| 4 | 6 | 4 | 6 | 1: 1.50 | | | | | | | | | | | | | | | 10 | 201-210 | 3 | 7 | 2.5.4
| 7.8 | 7.6 | {2.8.7.6 | 2 | 6.5.4 | 7.6 | 7 | 4 | 6 | 4 | 6 | 1: 1.50 | | | | | | |
{8.7.6.5 | | | | | | | | | | | | | | | | | | | | | | | | | 11 | 211-220 | 3 | 7 | 6.5.4
| 7.8 | 6 | {7.6.4.3 | 2 | 6.5.4 | 7.6 | 7 | 5 | 5 | 5 | 5 | 1: 1.00 | | | | | | |
{2.7.6.5 | | | | | | | | | | | | | | | | | | | | | | | | 12 | 221-230 | 3 | 7 | 2.4 |
7.8 | 6 | 7.6.5 | 2 | 6.5.4 | 8.7.6 | 7 | 5 | 5 | 5 | 5 | 1: 1.00 19 | 231-240 | 3 |
7 | 2.4 | 7.8 | 6 | 5 | 2 | 6.4 | 8.7.6 | 7 | 6 | 4 | 6 | 4 | 1: 0.67 | | | | | | | |
| | | | | | | 20 | 241-250 | 3 | 7 | 5.4 | 8 | 7.6 | {2.4.1.2.7 | 3.2 | 6.4 | 7.6 | 7
| 4 | 6 | 4 | 6 | 1: 1.50 | | | | | | {8.7.6.5 | | | | | | | | | | | | | | | | | | | |
| | | | | 21 | 251-260 | 3 | 7 | 6.5.4 | 7.8 | {7.4.5.4 | {6.4.3.2 | 2 | 6.5.4 |

8.7.6 | 7 | 4 | 6 | 4 | 6 | 1: 1.50 | | | | | | {8.7.6 | {7.6.5 | | | | | | | | | | | | |
| | | | | | | | | | | 23 | 261-270 | 3 | 7 | 6.5.4 | 7.8 | 6 | 6.5 | 2 | 6.5.4 | 7.6
| 7 | 5 | 5 | 5 | 5 | 1: 1.00 24 | 271-280 | 3 | 7 | 6.4 | 7.8 | 7.6 | 2.5 | 2 | 4 |
7.4.3.8.7.6 | 7 | 5 | 5 | 5 | 5 | 1: 1.00 25 | 281-290 | 3 | 7 | 2.5.4 | 8 | 7.6 | 5
| 2 | 6.4 | 7.6 | 7 | 6 | 4 | 6 | 4 | 1: 0.67 26 | 291-300 | 3 | 7 | 6.5.4 | 8 | 7.6
| 7.6.5 | 2 | 6.5.4 | 7.6 | 7 | 5 | 5 | 5 | 5 | 1: 1.00 27 | 301-310 | 3 | 7 |
2.6.5.4 | 8 | 7.6 | 5 | 2 | 6.5.4 | 7.5.4.9.8.6 | 7 | 6 | 4 | 6 | 4 | 1: 0.67 28 |
311-320 | 3 | 7 | 2.5.4 | 8 | 6 | 9.8.7.5 | 2 | 3.4 | 8.6 | 7 | 6 | 4 | 6 | 4 | 1:
0.67

=======+===========+==============+==============+============
==+=============+==============+==============+=============+
=============+==============+==============+===+===+===+===+==
======

[Footnote *: Aided by experimenter.]

I immediately set about reinstating the former attitude by lessening the number of trials and the punishment, and by increasing the value of the reward, but my best efforts, continuing up to August 28, failed markedly to improve the condition. The number of correct choices did somewhat increase, but at no time did the animal attain the degree of success which he had achieved as early as July 31 in the eleventh series of trials.

During the last two weeks of experimentation, all possible efforts were put forth to discover the best combination of rewards and punishments. Punishment was varied from 0 to confinement of sixty seconds, and many kinds of food in different amounts were tried as rewards, but in spite of everything Sobke failed to improve markedly. From time to time, notably on August 12 and 21, he exhibited peculiarly strong resentment toward me and repeatedly attempted to attack me.

The outcome of my experiments with problem 4 is peculiarly interesting in that it indicates the importance of a favorable attitude toward the work and the extreme risk from disturbing or discouraging conditions. It seems not improbable that had the work progressed without change in experimenter, or method of procedure, and above all without the disturbance of the painting,

Sobke might have solved problem 4 within a few days. This is by no means certain, however, for in problems 2 and 3 the ratio of right to wrong choices instead of increasing steadily increased very irregularly.

The detailed results for this problem are given in table 7. Reactive tendencies which appear are: (a) persistent choice of the end boxes followed, subsequently, by (b) the tendency to locate the middle box directly. This proved fairly easy when the number of boxes involved was only three as in settings 1, 4, 7, and 10. Setting 4 was most difficult of all, because box 9 was avoided or ignored. When the number of open boxes was as great as five, as in settings 2 and 8, the task was obviously more difficult, but whereas success in setting 2 appeared early, in setting 8 it failed to appear during the course of experimentation. For the settings 3, 6, and 9, involving either seven or nine open boxes, the direct choice of the middle box was next to impossible, and Sobke tended to choose, first of all, a particular box toward one end of the series, for example, box 2, in setting 3, and box 7 in setting 9. To the experimenter, as he watched the animal's behavior, it looked as though effort each time were being made to locate the middle member of the group. This appeared relatively easy for groups of three boxes, extremely difficult for as many as five boxes, and almost impossible for seven or nine.

3. Julius, Pongo pygmaeus _Problem 1. First at the Left End_

The orang utan, Julius, was gentle, docile, and friendly with the experimenter throughout the period of investigation. He at no time showed inclination to bite and could be handled safely. As contrasted with Skirrl and even with Sobke, he adapted himself to the multiple-choice apparatus very promptly, and only slight effort on the part of the observer was necessary to prepare him, by preliminary trials, for the regular experiments. But in order to facilitate work, he was familiarized with the apparatus by means of regular route training and feeding in the several boxes from April 5 to April 9.

On April 10 the apparatus was painted white as has been stated previously, and on the following Monday, April 12, Julius when again introduced to it gave no indications of fear, uneasiness, or dislike, but worked as formerly, making his round trips quickly and eagerly entering any box which happened to be open, in order to obtain the reward of food.

The regular experimentation was undertaken on April 13, and the results of the first series of trials with Julius are sharply contrasted with those obtained with the monkeys in that fewer choices were necessary. Instead of the expected ratio of right to wrong first choices, 1 to 2.5, the orang utan gave a ratio of 1 to 1. An additional markedly different result from that obtained with the monkeys is indicated below in the total time required for a series of trials. As examples, the data for the first, second, fifth, and tenth series are presented.

TIME FOR SERIES OF TRIALS

1st series 2nd series 5th series 10th series Skirrl 35 min. 20 min. 14 min. 10 min. Sobke 14 " 17 " 10 " 9 " (8th series) Julius 12 " 11 " 14 " 9 "

It is also noteworthy that Julius in the presence of visitors or under other unusual conditions worked steadily and well, whereas the monkeys, and especially Sobke, tended to be distracted and often refused to work at all.

Almost from the beginning of his work on problem I, Julius began to develop the tendency to enter immediately the open door nearest the starting point. In case the group of open doors lay to the right of the middle of the apparatus, this method naturally yielded success; whereas if the group included doors to the left of the middle, it resulted in failure. Obviously it was a most unsatisfactory method, and although it enabled him to make more right than wrong first choices, it prevented him from increasing the number of right choices, and as table 1 indicates, it maintained the ratio of 1 right to .67 wrong first choices for eight successive days.

On April 23 a break occurred in which the number of correct choices was reduced from six to five. Julius worked very rapidly and with almost no hesitation in choosing. My notes record "he seems to miss the point wholly. It is doubtful whether the punishment is sufficiently severe." At this time he was being punished by thirty seconds confinement in each wrong box, the interval having been held fairly steadily from the first series of experiments. On April 26 it was increased to sixty seconds, in an effort to break him of the habit of choosing the "nearest" door. But he became extremely restless under the longer confinement and tried his best to raise the entrance and exit doors. Since there was at this time no mechanism for locking them when closed, it was difficult for the experimenter to prevent him from escaping by way of the entrance door or from raising the exit door sufficiently to obtain the food. Indeed, the longer confinement worked so unsatisfactorily that on the following day I substituted for it the punishment of forcing him to raise the entrance door of the wrong box in order to escape for a new choice. He was rewarded with food in the alleyway H, beside door 15 (figure 17), only when he chose correctly on first attempt.

This method discouraged him extremely and proved wasteful of time. Consequently, in a second series on the same date return was made to the former method, and he was rewarded with food whenever he found the right box. But on April 28, the two methods were again employed, the first in the initial series and the second in a final series of trials. The animal's persistent attempts to raise the doors gave the experimenter so much trouble that on April 29 barbed wire was nailed over the windows of the entrance doors with the hope that it might prevent him from working at them. But he quickly learned to place his fingers between the barbs and raise the doors as effectively as ever.

On April 30 the reward of food was given only when the first choice was that of the right box and in that event it was placed in the alleyway H as stated above.

As it seemed absolutely essential to break the unprofitable habit of choosing the nearest door, on May 3 a new series of settings was presented, in which only the doors to the left of the middle of the row of nine boxes were used as right doors. That is, in this new series, doors 1 to 4 occur as right doors; 5 to 9 do not. As punishment for wrong choices on this date, Julius was confined in the wrong box from one to five minutes. It was difficult to keep him in, but by means of cords which had been attached to the doors, this was successfully accomplished. Yet another and slightly different series of settings was employed on May 4, and this, proving satisfactory, was continued in use until the end of the experiment, with punishment ranging from sixty to one hundred and twenty seconds for each mistake.

Naturally the modification of settings introduced May 3 greatly increased the proportion of wrong first choices. Indeed, as appears in table 8, the ratio of right to wrong immediately changed from 1:0.67 to 1:4.00. Between May 3 and May 10, no steady and consistent improvement in method or in the number of correct first choices occurred, and on the last named date, Julius chose correctly only three times in his ten trials. At this time there was, as my notes record, no satisfactory indication of progress, and the status of the experiment seemed extremely unsatisfactory in as much as in spite of the experimenter's best efforts to break up the habit of choosing the nearest door, the orang utan still persisted, to a considerable extent, in the use of this method. The only encouraging feature of the results was an evident tendency to choose somewhat nearer the left end of a group than previously.

TABLE 8

Results for Orang utan in Problem 1

========+============+==============+==============+============
==+=============+==============+==============+==============+
=============+==============+==============+===+===+===+===+==
====== | No. | S.1 | S.2 | S.3 | S.4 | S.5 | S.6 | S.7 | S.8 | S.9 | S.10 | | | | |
Ratio Date | of | | | | | | | | | | | R | W | R | W | of | trials | 1.2.3 | 8.9 |
3.4.5.6.7 | 7.8.9 | 2.3.4.5.6 | 6.7.8 | 5.6.7 | 4.5.6.7.8 | 7.8.9 | 1.2.3 | | | | | R
to W --------+------------+--------------+--------------+--------------+---------------+---------------+------------
---+---------------+---------------+---------------+---------------+---------------+---+---+---+---+--
------ | | | | | | | | | | | | | | | | April | | | | | | | | | | | | | | | | 13 | 1- 10 |
3.1 | 8 | 4.3 | 7 | 4.2 | 7.6 | 5 | 4 | 7 | 3.1 | 5 | 5 | 5 | 5 | 1:1.00 14 | 11- 20 |
3.2.1 | 8 | 4.3 | 7 | 4.4.2 | 6 | 5 | 4 | 7 | 3.1 | 6 | 4 | 6 | 4 | 1:0.67 15 | 21-
30 | 3.2.1 | 8 | 4.3 | 7 | 4.5.5.2 | 6 | 5 | 4 | 7 | 3.1 | 6 | 4 | 6 | 4 | 1:0.67 16 |
31- 40 | 3.1 | 8 | 4.3 | 7 | 4.2 | 6 | 5 | 4 | 7 | 3.2.1 | 6 | 4 | 6 | 4 | 1:0.67 17 |
41- 50 | 3.2.1 | 8 | 4.3 | 7 | 4.2 | 6 | 5 | 4 | 7 | 3.1 | 6 | 4 | 6 | 4 | 1:0.67 19 |
51- 60 | 3.1 | 8 | 4.3 | 7 | 4.2 | 6 | 5 | 4 | 7 | 3.1 | 6 | 4 | 6 | 4 | 1:0.67 20 |
61- 70 | 2.1 | 8 | 4.3 | 7 | 5.3.2 | 6 | 5 | 4 | 7 | 3.1 | 6 | 4 | 6 | 4 | 1:0.67 21 |
71- 80 | 3.1 | 8 | 4.3 | 7 | 5.4.3.2 | 6 | 5 | 4 | 7 | 3.2.1 | 6 | 4 | 6 | 4 | 1:0.67
22 | 81- 90 | 3.1 | 8 | 5.3 | 7 | 6.3.2 | 6 | 5 | 4 | 7 | 3.2.1 | 6 | 4 | 6 | 4 |
1:0.67 23 | 91-100 | 3.2.1 | 8 | 5.3 | 7 | 4.3.2 | 6 | 5 | 5.4 | 7 | 3.2.1 | 5 | 5 |
5 | 5 | 1:1.00 24 | 101-110 | 3.2.1 | 8 | 4.3 | 7 | 4.3.2 | 6 | 5 | 4 | 7 | 3.1 | 6
| 4 | 6 | 4 | 1:0.67 26 | 111-120 | 3.1 | 8 | 4.3 | 7 | 4.3.2 | 6 | 5 | 5.4 | 7 |
3.1 | 5 | 5 | 5 | 5 | 1:1.00 27 | 121-130 | 3.2.1 | 8 | 4.3 | 7 | 4.3.2 | 6 | 6.5 |
5.8.6.4 | 7 | 3.3.3.1 | 4 | 6 | | | " | 131-140 | 3.1 | 8 | 4.3 | 7 | 4.3.2 | 6 | 5 |

4 | 7 | 3.2.1 | 6 | 4 |10 |10 | 1:1.00 28 | 141-150 | 3.2.1 | 8 | 3 | 7 | 5.4.2 | 6 | 5 | 4 | 7 | 3.1 | 7 | 3 | | | " | 151-160 | 3.1 | 8 | 3 | 7 | 3.2 | 6 | 5 | 4 | 7 | 3.2.1 | 7 | 3 |14 | 6 | 1:0.43 29 | 161-170 | 3.1 | 8 | 4.3 | 7 | 4.3.2 | 6 | 5 | 4 | 7 | 3.2.1 | 6 | 4 | | | " | 171-180 | 3.2.1 | 8 | 4.3 | 7 | 4.2 | 6 | 5 | 4 | 7 | 3.2.1 | 6 | 4 |12 | 8 | 1:0.67 | | | | | | | | | | | | | | | | | | | | | 30 | 181-190 | 3.1 | 8 | 4.3 | 7 | {4.5.6.4 | 6 | 5 | 4 | 7 | 3.1 | 6 | 4 | | | | | | | | | {5.6.4.2 | | | | | | | | | | | | | | | | | | | | | | | | | | | | | " | 191-200 | 3.1 | 8 | 4.5.6.7.3 | 7 | 4.5.3.2 | 6 | 5 | 4 | 7 | 3.2.1 | 6 | 4 |12 | 8 | 1:0.67 May | | | | | | | | | | | | | | | | | 1 | 201-210 | 3.1 | 8 | 4.3 | 7 | 3.2 | 6 | 5 | 4 | 7 | 3.1 | 6 | 4 | | | " | 211-220 | 3.2.1 | 8 | 4.3 | 7 | 4.2 | 6 | 5 | 4 | 7 | 3.1 | 6 | 4 |12 | 8 | 1:0.67 --
-------+----------+--------------+--------------+--------------+--------------+--------------+-----
---------+--------------+--------------+--------------+--------------+--------------+---+---+---+---+-------- |
| | | 2.3.4.5 | | | | | | | 2.3.4.5 | | | | | | | 1.2.3 | 3.4.5.6.7 | 6.7.8 | 1.2.3 | 3.4.5.6.7 | 4.5.6.7.8.9 | 2.3.4.5 | 1.2.3 | 4.5.6.7.8.9 | 6.7.8.9 | | | | | --------+--
----------+--------------+--------------+--------------+--------------+--------------+--------------
+--------------+--------------+--------------+--------------+--------------+---+---+---+---+-------- 3 | 221-
230 | 3.1 | 4.3 | 4.2 | 3.1 | 4.3 | 4 | 4.3.2 | 3.1 | 4 | 4.3.2 | 2 | 8 | 2 | 8 | 1:4.00 --------+----------+--------------+--------------+--------------+--------------+--------------+----------
-----+--------------+--------------+--------------+--------------+--------------+--------------+---+---+---+---
+-------- | | | | 2.3.4.5 | 4.5.6.7 | | | | | 2.3.4.5 | | | | | | | 1.2.3 | 3.4.5.6.7 | 6.7.8 | 8.9 | 2.3.4.5 | 3.4.5.6 | 1.2.3 | 4.5.6.7.8 | 6.7.8.9 | 1.2.3 | | | | | -----
---+----------+--------------+--------------+--------------+--------------+--------------+--------
-------+--------------+--------------+--------------+--------------+--------------+---+---+---+---+-------- 4 |
231-240 | 3.2.1 | 4.3 | 4.2 | 4 | 4.3.2 | 4.3 | 3.2.1 | 4 | 3.2 | 3.2.1 | 2 | 8 | 2 | 8 | 1:4.00 5 | 241-250 | 2.1 | 3 | 3.2 | 4 | 3.2 | 4.3 | 3.2.1 | 4 | 3.2 | 3.2.1 | 3 | 7 | 3 | 7 | 1:2.33 6 | 251-260 | 2.1 | 3 | 2 | 4 | 3.2 | 3 | 2.1 | 4 | 3.2 | 2.1 | 5 | 5 | 5 | 5 | 1:1.00 7 | 261-270 | 2.1 | 3 | 3.2 | 4 | 3.2 | 3 | 2.1 | 4 | 4.2 | 2.1 | 4 | 6 | 4 | 6 | 1:1.50 8 | 271-280 | 2.1 | 4.3 | 4.3.2 | 4 | 3.2 | 4.3 | 3.1 | 4 | 3.2 | 2.1 | 2 | 8 | 2 | 8 | 1:4.00 10 | 281-290 | 1 | 4.3 | 4.2 | 4 | 3.2 | 4.3 | 2.1 | 4 | 3.2 | 2.1 | 3 | 7 | 3 | 7 | 1:2.33 11 | 291-300 | 1 | 3 | 2 | 4 | 2 | 3 | 1 | 4 | 2 | 1 |10 | 0 |10 | 0 | 1:0.00 --------+----------+--------------+--------------+--------------+-
--------------+--------------+--------------+--------------+--------------+--------------+--------------+----------
----+--------------+---+---+---+---+-------- | | | | | | | | | | | | | | | | | | | | | | | 1.2.3 | 8.9 | 3.4.5.6.7 | 7.8.9 | 2.3.4.5.6 | 6.7.8 | 5.6.7 | 4.5.6.7.8 | 7.8.9 | 1.2.3 | | | |
--------+----------+--------------+--------------+--------------+--------------+--------------+---
-----------+--------------+--------------+--------------+--------------+---+---+---+---+--------
12 | 301-310 | 1 | 8 | 3 | 7 | 2 | 6 | 5 | 4 | 7 | 1 |10 | 0 |10 | 0 | 1:0.00
========+==========+==============+==============+==============+==============

==+==============+==============+==============+==============+
==============+==============+==============+===+===+===+===+==
======

A series of correct first choices was obtained on May 11, greatly to the surprise of the experimenter, for no indication had previously appeared of this approaching solution of the problem. It seemed possible, however, that the successes were accidental, and it was anticipated that in a control series Julius would again make mistakes. But on the following day, May 12, the presentation of the original series of ten settings, which, of course, differed radically from the settings used from May 4 to May 11 was responded to promptly, readily, and without a single mistake. Julius had solved his problem suddenly and, in all probability, ideationally.

Only three reactive tendencies or methods appeared during Julius's work on this problem: (a) choice of the open door nearest to the starting point (sometimes the adjacent boxes were entered); (b) a tendency to avoid the "nearest" door and select instead one further toward the left end of the group; (c) direct choice of the first door on the left.

The curve of learning plotted from the daily wrong choices and presented in figure 18, had it been obtained with a human subject, would undoubtedly be described as an ideational, and possibly even as a rational curve; for its sudden drop from near the maximum to the base line strongly suggests, if it does not actually prove, insight.

Never before has a curve of learning like this been obtained from an infrahuman animal. I feel wholly justified in concluding from the evidences at hand, which have been presented as adequately as is possible without going into minutely detailed description, that the orang utan solved this simple problem ideationally. As a matter of fact, for the solution he required about four times the number of trials which Sobke required and twice as many as were necessary for Skirrl. Were we to measure the intelligence of these three animals by the number of trials needed in problem 1, Sobke clearly would rank first, Skirrl second, and Julius last of all. But other facts clearly indicate that Julius is far superior to the monkeys in intelligence. We therefore must conclude that _where very different methods of learning appear, the number of trials is not a safe criterion of intelligence._ The importance of this conclusion for comparative and genetic psychology needs no emphasis.

Problem 2. Second from the Right End

Julius was given four days' rest before being presented with problem 2. He was occasionally fed in the apparatus, but regular continuation of training was not necessary to keep him in good form. During this rest interval, locks were attached to the doors of the apparatus so that the experimenter by moving a lever directly in front of him could fasten either one or both of the doors of a given box by a single movement. On May 13 Julius was given opportunity to obtain food from each of the boxes in turn, and trial of the locks was made in order to familiarize him with the new situation. He very quickly discovered that the doors could not be raised when closed, and after two days of preliminary work, he practically abandoned his formerly persistent efforts to open them. The locks worked satisfactorily from a mechanical point of view as well as from that of the adaptation of the animal to the modified situation.

Problem 2 was regularly presented for the first time on May 17, on which day a single series was given. The period of punishment adopted was twenty seconds, and for each successful choice a small piece of banana was given as a reward. After the first trial in this series, in which Julius repeatedly entered the first box at the left, that is box 7, there was but slight tendency to reenter the first box at the left of the group. Instead, Julius developed the method of moving box by box toward the right end of the group. The choices were made promptly, and their systematic character enabled the animal to obtain his reward fairly quickly, in spite of the large number of mistakes.

In the second series, the orang utan developed the interesting trick of quickly dodging out of the wrong box before the experimenter could lower the door behind him. This he did only after having been punished for many wrong choices to the point of discouragement. The trick was easily broken up by the sudden lowering of the entrance door as soon as he had passed under it.

There appeared on May 21 an unfavorable physical condition which manifested itself, first of all through the eyes which appeared dull and bloodshot. On the following day they were inflamed and the lids nearly closed. Julius refused to eat, and experimentation was impossible. Until June 2 careful treatment and regulation of diet was necessary. He passed through what at the time seemed a rather startling condition, but rapidly regained his usual good health, and on June 3, although somewhat weak and listless, he again worked fairly steadily.

Since it was now possible to lock the doors and confine the animal for any desired period, on June 5 the interval of punishment was made sixty seconds, and a liberal quantity of banana, beet, or carrot was offered as reward. No increase in the number of successful choices appeared, and Julius showed discouragement. Sawdust had been strewn on the floor, and in the intervals between trials as well as during confinement in wrong boxes, he took to playing with the sawdust. He would take it up in one hand and pour it from hand to hand until all had slipped through his fingers, then he would scrape together another handful and go through the same process. Often he became so intent on this form of amusement that even when the exit door was raised, he would not immediately go to get the food.

The reactive tendencies which appeared in the work on problem 2 will now be presented in order, since I shall have to refer to them repeatedly, and the list will be more useful to the reader at this point than at the conclusion of the presentation of daily results. The following is not an exhaustive list but includes only the most important and conspicuous tendencies or methods together with the dates on which they were most apparent.

(a) May 17, choice of first box at left of group or near it, then the next in order, and so on, until the second from the right was reached. This method with irregularities and certain definite skipping was used at various times, sometimes over periods of several days, during the course of the work.

(b) June 3, preference for number 3 and number 4 developed immediately after the orang utan's illness and when he was working rather listlessly.

On June 9 and 10, the original tendency (a) reappeared and persisted for a number of series.

(c) June 14, a tendency to choose the box at or near the right end of a group, and then the one next to it. In connection with this tendency, which of course required only two choices in any given trial, interest in playing with the sawdust on the floor developed.

Again on June 21, the animal returned to the use of tendency (a).

(d) June 29, movement to box at right end of group, hesitation before it, and turning through a complete circle so that the second box from the right was faced. This, the correct box, was often promptly entered. This method, if persisted in, would obviously have yielded solution of the problem.

(e) July 5, approach to and pretense to enter the box next to the right end (right one), and then choice of some other box. This feint is peculiarly interesting, and its origin and persistence are difficult to account for.

(f) In connection with the tendency to pretend that he was going to enter the second box from the right end, Julius developed also the tendency to turn around in front of the box at the right end, starting sometimes to back into it, and then to enter, instead, the box second from the end.

(g) July, 6 and 7, a fairly definite tendency to take the one next in order or, instead, to go directly to the right box.

(h) July 10, direct first choices without approach to other boxes appeared for the first time on this date.

For this problem, it proved impossible to establish and maintain uniform conditions of experimentation. Instead, because of the failure of the animal to improve and the tendency to discouragement, both punishment and reward had to be altered from time to time, and other and more radical changes were occasionally made in the experimental procedure. Below for the sake of condensed and consecutive presentation, the most important conditions from day to day are arranged in tabular form:

CONDITIONS OF EXPERIMENT FROM DAY To DAY FOR PROBLEM 2

Date Punishment Reward

May 17 .... 20 sec. confinement .... Food in right box for each (Aid after 10 trials) trial

" 18 to 21 .... 30 sec. confinement .... Food (banana) in right box for each trial

" 22 to June 2 .... Illness, no experiments

June 3 .... 15 sec. confinement .... Food (banana) in right box for each trial

" 4 .... 30 " " .... Food (banana) in right box for each trial

" 5-10 .... 60 " " .... Beet, carrot and loquat, in addition to banana

" 11 .... 10 to 30 sec. confinement .. Beet, carrot and loquat, in addition to banana

" 12 to 15 .... 60 sec. confinement .... Beet, carrot and loquat, in addition to banana

" 16 .... 60 " " .... Banana and sweet corn--former preferred

" 17 (1st series). 60 sec. confinement .... Food (banana, as in early series)

" 17 (2nd series). No confinement in wrong box; Food only for correct first but instead, return to choices starting point by way of alleys

" 18 to 22 .... No confinement in wrong box; Food only for correct first but instead, return to choices starting point by way of alleys

" 22 (2nd series). No punishment; allowed to Food for each trial enter boxes until right one was found " 23 .... Return to starting point. After five wrong choices of a given box the animal was held for 60 secs. in one of the boxes and was then released by way of the exit door and rewarded when the right one was chosen

" 23 (2nd series). No punishment .... Reward for each trial

" 24 (1st series). Return to starting point. .. Food only for correct first choices

" 24 (2nd series). No punishment .... Reward for each trial

" 25-30 .... Same as on 24th ....

July 1 (1st series). No punishment .... " " " "

" 1 (2nd series). Return to starting point ... Reward only for correct first choices " 2-8 .... Same as on July 1 ....

" 8 (2nd series). No punishment .... Reward for each trial

" 8 (3rd series). Return to starting point ... Reward only for correct first choices

" 9-10 .... Same as for July 8 (3rd series)

" 10 (2nd series). Momentary confinement in ..... Reward for each correct choice wrong boxes

" 12 .... Return to starting point ..... Reward for correct first choice

" 12 (2nd series). 30 sec. confinement .... Reward for each correct choice

" 12 (3rd series). 5 " " .... " " " " "

" 13 .... 30 " " .... " " " " "

" 14-17 .... Return to starting point ..... Reward for correct first choices

" 17 (2nd series). 60 sec. confinement .... Reward for each correct choice

" 19 .... 30 " " .... " " " " "

" 20-26 .... 10 " " .... " " " " "

" 27-30 .... Right box indicated by slight Reward in each right box raising of exit door momentarily. No punishment

" 30 (2nd series). Return to starting point ..... Reward for correct first choices

" 31 .... " " " " ..... " " " " "

" 31 (2nd series) to Aug. 10 .... 10 to 60 sec. confinement .... Reward for each correct choice

Aug. 10 (2nd series). Threatened with whip .... " " " " "

" 11 (1st series). " " " .... " " " " "

" 11 (2nd series). 10 sec. confinement .... " " " " "

" 12 .... Threatened with whip .... " " " " "

" 12 (2nd series). 10 sec. confinement .... " " " " "

" 19 .... 10 " " .... " " " " "

" 19 (2nd series). Threatened with whip .... " " " " "

With the above reactive tendencies and modifications of method in mind we may continue our description of results. On June 9 there developed a tendency to increase the magnitude of the original error by choosing nearer the left end of the groups. This is odd, since one would naturally suppose that an animal as intelligent as the orang utan would tend to avoid the general region in which success was never obtained and to focus attention on the right, as contrasted with the wrong end of each group. _It obviously contradicts the law of the gradual elimination of use less activities._ In other words, it is wholly at variance with the principle of trial and error exhibited by many infrahuman organisms. Julius, although making many mistakes, worked diligently and, for the most part, fairly rapidly. The day's work proved most important because of the change in method and also because of the appearance of hesitation, the rejection of certain boxes, and the definite choice of others. My notes record "this is a most important day for Julius in problem 2;" but subsequent results do not clearly justify this prophecy.

The method of choosing the first box at the left and then of moving down the line until the right one was reached was so consistently followed that during a number of days it was possible for me to predict almost every choice. Indeed, to satisfy my curiosity in this matter during a number of series I guessed in advance the box which would be chosen. The percentages of correct guesses ranged from ninety to one hundred. June 10, for example, yielded two series for which the ratio of right to wrong first choices was 0 to 10, and in which the method described above was used consistently throughout.

It was inevitable that punishment by confinement and the discouragement resulting therefrom should interfere with the regularity of work and make it extremely difficult to obtain strictly comparable results from series to series and from day to day. The data for this problem, as presented in table 9, have values quite different from those for the monkeys, chiefly because of the more variable conditions of observation.

It was occasionally noted that the disintegration of a definite method and the disappearance of the tendency on which it depended occurred rather suddenly. Frequently it happened that having used an inadequate method fairly persistently on a given day, the animal would on the following day exhibit a wholly different method. Even over night a new method might develop. In the monkeys, although there was occasionally something comparable with this, it was by no means so evident.

After two hundred and fifty trials on problem 2 had been given Julius, it seemed desirable to introduce a radical change in method in order to stimulate him to maximal effort. It was therefore decided to force him to make a round trip through the apparatus in connection with each choice, and to let this forced labor serve, in the place of confinement, as punishment for mistakes. This new method yielded peculiar and characteristic results. They differ from those previously obtained largely because of the orang utan's remarkably strong tendency to reenter the box through which he had just passed. This occurred so persistently, as may be seen in table 9 (June 17, second series, June 18, etc.), that a further modification of method was introduced in that after the same wrong box had been entered five times in succession, the experimenter on the next choice of the box confined the animal for a stated interval, say sixty seconds, in it, and then allowed it to escape by way of the exit door and choose repeatedly until it finally located the right box. Were it not for this particular feature of the method, the number of choices recorded after June 17 would unquestionably be very much greater than the table indicates.

The new method proved a severe test of the orang utan's patience and perseverance, for he had to work much harder than formerly for his reward, and often became much fatigued before completing the regular series of ten trials. Early in the use of this method, he developed the habit of rolling around from exit door to starting point by a series of somersaults. When especially discouraged he would often bump his head against the floor so hard that I could hear the dull thud. As has been noted, I found it desirable to vary the procedure repeatedly. It proved especially interesting to give one series per day with the round trip as punishment and another series with confinement as punishment.

Day after day, as the experiment progressed, slight or great fluctuations of the ratios of right to wrong choices appeared, but without consistent improvement. There was, to be sure, as the last column of table 9 shows, a radical improvement during the first six hundred and fifty trials, for the number of right choices per series increased from 0 to 8. But, as the observations were continued from day to day, it became more and more evident that the animal was merely passing from tendency to tendency--method to method--mixing tendencies, and occasionally developing new ones, without approach to the solution of the problem. This fact would have led me to discontinue the work much earlier than I actually did had it not been for the peculiarity of the results obtained with problem 1. It seemed not improbable that at any time Julius might succeed in perfectly solving this problem over night precisely as he had solved the first problem.

A curiously interesting bit of behavior appeared for the first time on June 29. Julius had gone to the first box at the right end of the group, and instead of entering, he had wheeled around toward his right, and turning a complete circle, faced the right box, which he promptly entered. Subsequently, the tendency developed and the method was used with increasing frequency. On June 30, it appeared in the first series, four times, in the second series, six times; on July 1, in the first series, three times, and in the second series, four times; on July 2, in the first series, five times, and in the second series, nine times. It was indeed only by accident that the animal failed to fulfill the technical requirement for perfect solution of the problem in this series. Yet, had he done so, his subsequent trials would doubtless have revealed the lack of any other idea than that of turning completely around before entering a box.

This odd bit of behavior proved peculiarly interesting and significant in that the tendency to turn became dissociated from the position (in front of the first box at the right end of the group) in connection with which it originally developed. After a few days, Julius would enter the reaction-chamber and instead of proceeding directly to the right end of the group, would stop suddenly wherever he happened to be, turn toward his right in a complete circle, and hasten into the box nearest to him which, as often as not, proved to be the wrong one. Thus the idea of turning completely about, which had it continued its association with the idea of facing the first box at the right, would have yielded success, instead became useless because of its dissociation. That the orang utan is capable of using free ideas seems clear enough in the light of this behavior. That he proved incapable of getting the idea of second from the right end is as clearly shown by the detailed results of table 9,--the fruits of weeks of experimenting.

Certain other interesting tricks developed in Julius's behavior. Thus, on July 5, there appeared the tendency to move as though about to enter the right box (feint), then to stop suddenly and promptly enter another box, which was, of course, a wrong one. The reason for the development of this tendency could not be discovered, but in connection with it, there appeared another tendency which possibly can be explained. Julius took to backing into the chosen box so that he could face the experimenter. He would then, after a period of hesitation, come out and promptly enter one of the other boxes. This tendency was apparently due to the fact that during one or two series the experimenter growled at the orang utan every time he made a mistake. The growl startled him and caused him to look around. He evidently felt the need of keeping his eyes on the experimenter,--Hence the backing into the open box. The tendency disappeared shortly after the experimenter gave up the use of the growl as a method of punishing the animal for what were suspected to be careless choices.

Curiously enough, it was not until July 10 that direct choice of the right box was made at all frequently. Previously, selection of it had been made almost invariably after approach to other boxes. But in the second series for July 10 there was an extraordinary improvement in method. This developed in the presence of two visitors, and it is therefore all the more surprising. The choices were made not only directly, but with decision and evident certainty that was quite at variance with the previous behavior of the animal.

All the while through variation of methods, I was seeking to discover the best means of holding the orang utan to his maximum effort and care in attempting to select the right box. One day it would seem as though forcing him to make round trips with rewards only for correct first choices proved most satisfactory, and the next it might seem equally clear that punishment by confinement for thirty seconds or sixty seconds, with reward for correct choice in every trial, yielded better results. In the end I had to admit that no best method had been demonstrated and that I had failed to develop conditions which served to compel the animal's attention to the problem and to lead him to work without discouragement. There were, it is true, days on which it seemed practically certain that the problem would be solved, but as it turned out, Julius never succeeded in choosing correctly--throughout a series of ten trials.

As a last resort, in order to make perfectly sure that the orang utan was doing his best, I decided to introduce corporal punishment in a mild form. For this purpose, I placed my assistant in charge of the apparatus and the series of trials, and stationed myself in one corner of the reaction-chamber with a whip in my hand. Whenever Julius entered a wrong box, I approached him with the whip and struck at him, being careful not to injure him and rarely striking him at all, for the threat was more effective than a blow. He was extremely afraid of the whip and would begin to whine and attempt to get out of the way as soon as he saw it.

This method was introduced on August 10, but no improvement resulted, and in the end there was no reason to consider it more satisfactory than the other procedures. I am now wholly convinced that Julius did his best to choose correctly in the majority of the numerous series which were given him in connection with problem 2.

From trials 1001 to 1100, a radical departure from the previous methods was introduced in that the right box was indicated to the animal by the slight and momentary raising of its exit door. Of course no records of the choices for this group of one hundred trials appear in table 9, for the simple reason that the animal inevitably and immediately entered the right box. It was thought that this method might serve to break up the previously developed tendencies toward inadequate forms of response and so encourage the animal that he would later solve the problem when given opportunity to select the right box without aid from the experimenter. But as a matter of fact, while the ratio of right to wrong first choices was 1 to .67 in the series preceding this change of method, it was 1 to 1.50 in the first series following its use. There is no satisfactory evidence that Julius profited by this experience, though as a matter of fact he did succeed in making his best daily record, eight right to two wrong choices, on August 4, after 1190 trials.

The curve of learning for this problem has been plotted and is presented in figure 19. It is of course incomplete and it is offered only to indicate the extreme irregularity in performance.

Problem 1a. First at the Right End

It was decided on August 19 that the further continuation of the work of Julius on problem 2 was not worth while. He had become much discouraged, and although willing to work for food, gave no indications whatever of improvement and seemed to have exhausted his methods. It seemed wise instead of giving up work with him in the multiple-choice method to return to a form of problem 1. We may designate it as problem 1a. The right box is definable as the first at the right end of the series instead of the first at the left end as in the original problem 1. It was thought possible that Julius might quickly solve this problem by a process similar to that used for problem 1.

Work was begun on problem 1a, August 20, and for six successive days two series of trials per day were given, the settings for which as well as the resulting choices are given in table 10. Most notable in these results is the large number of cases in which Julius chose first the second box from the right end of the series, or in other words that box which had been the right one in problem 2. Contrary to expectation, he showed no inclination to abandon this tendency to choose the second from the right end, and the ratio of right to wrong choices changed in the direction opposite from expectation, beginning with 1 to 4 and ending on the sixth day with 0 to 20.

It was obviously useless to continue the experiment further since Julius had given up his attempts to locate the right box in the first choice and was apparently satisfied to discover it by a process of trial and error. He had, it would seem, satisfied himself that the problem was insoluble. These results obtained in problem 1a constitute a most interesting comment on the effects of problem 2 on the orang utan. Behavior similar to that which he developed well might have been obtained from a child of three to four years placed in a like situation and forced to strive, day after day, to solve a problem beyond its ideational capacity.

In many respects the most interesting and to the experimenter the most surprising result of this long series of observations with Julius was the lack of consistent improvement. It seemed almost incredible that he should continue, day after day, to make incorrect choices in a particular setting while choosing correctly in some other setting which from the standpoint of the experimenter was not more difficult.

TABLE 10

Results for Orang utan in problem 1a

```
========+===========+==============+==============+============
==+============+==============+==============+==============+
=============+==============+==============+===+===+===+===+==
====== | No. | S.1 | S.2 | S.3 | S.4 | S.5 | S.6 | S.7 | S.8 | S.9 | S.10 | | | |
Ratio Date | of | | | | | | | | | | 1.2.3.4 | R | W | R | W | of | trials | 5.63 |
1.2.3.4 | 6.7.8.9 | 2.3.4.5 | 3.4.5.6.7 | 1.2.3 | 5.6.7.8 | 1.2 | 2.3.4.5.6 | 5.6.7 |
| | | | R to W --------+-----------+--------------+--------------+--------------
+--------------+--------------+--------------+--------------+--------------+---+---
+---+---+-------- | | | | | | | | | | | | | | | | August | | | | | | | | | | | | | | |
20 | 1- 10 | 6 | 3.4 | 6.7.8.9 | 4.5 | 6.7 | 3 | 7.8 | 2 | 5.5.6 | 6.7 | 3 | 7 | | | |
| | | | | | | | | | | | | | | | | " | 11- 20 | 5.6 | 3.4 | {7.8.7.8 | 4.5 | 6.7 | 2.3 |
{7.6.7.7 | 2 | 5.6 | 6.7 | 1 | 9 | 4 |16 | 1: 4.00 | | | | {8.7.8.9 | | | | {6.7.7.8 |
| | | | | | | | | | | | | | | | | | | | | | | | 21 | 21- 30 | 5.6 | 3.4 | {7.8.7.6 | 4.5 |
```

6.7 | 2.3 | 7.8 | 2 | 5.6 | 5.7 | 1 | 9 | | | | | | | {8.7.9 | | | | | | | | | | | | |
| | | | | | | | | | | | | | | " | 31- 40 | 5.6 | 3.4 | 7.7.6.8.9 | 4.5 | 6.7 | 3 | 6.7.8
| 2 | 6 | 6.7 | 3 | 7 | 4 |16 | 1: 4.00 23 | 41- 50 | 5.6 | 3.4 | 7.8.9 | 4.5 | 6.7 |
2.3 | 6.7.8 | 2 | 5.6 | 5.6.7 | 1 | 9 | | | " | 51- 60 | 5.6 | 3.4 | 7.8.9 | 4.5 | 6.7
| 2.3 | 6.8 | 2 | 5.6 | 6.7 | 1 | 9 | 2 |18 | 1: 9.00 24 | 61- 70 | 5.6 | 3.4 | 6.8.9
| 4.5 | 5.7 | 2.3 | 6.7.8 | 1.2 | 5.6 | 6.7 | 0 |10 | | | " | 71- 80 | 5.6 | 3.4 |
6.7.8.9 | 4.5 | 5.7 | 2.3 | 5.7.8 | 2 | 5.6 | 6.7 | 1 | 9 | 1 |19 | 1:19.00 25 | 81-
90 | 5.6 | 3.4 | 6.7.8.9 | 5 | 5.6.7 | 2.3 | 7.8 | 1.2 | 5.6 | 4.5.6.7 | 1 | 9 | | | "
| 91-100 | 5.6 | 3.4 | 6.7.8.9 | 3.4.5 | 6.6.7 | 2.3 | 6.7.8 | 1.2 | 5.6 | 6.7 | 0
|10 | 1 |19 | 1:19.00 | | | | | | | | | | | | | | | | | 26 | 101-110 | 5.6 | 3.4 |
{6.7.8.8 | 3.5 | 5.6.7 | 2.3 | 5.6.7.6.7.8 | 1.2 | 5.6 | 6.7 | 0 |10 | | | | | | |
{6.7.6.9 | | | | | | | | | | | | | | | | | | | | | | | | | | | | | | | | {6.7.8.8 | | | |
| | | | | | | " | 111-120 | 5.6 | 2.3.4 | {6.7.8.7 | 3.4.5 | 5.6.7 | 2.3 | 7.8 |
1.2 | 5.6 | 4.5.6.7 | 0 |10 | 0 |20 | 0:20.00 | | | | {6.7.9* | | | | | | | | | | | |
| | | | | | | | | | | | | | | |
=========+=============+===============+===============+=============
==+=============+===============+===============+===============+
==============+===============+===============+===+===+===+===+==
======

[Footnote *: Aided by experimenter.]

The evidence suggests that in this young orang utan ideational learning tended to replace the simpler mode of problem solution by trial and error. Seemingly incapable of solving his problems by the lower grade process, he strove persistently, and often vainly, to gain insight. He used ideas ineffectively. Animals far lower in intelligence (e.g., the pig), surpass him in ability to solve these relational problems because they use the method of elimination by trial consistently and effectively. Julius, in these experiments, made a poor showing because his substitute for trial and error is only slightly developed. Would he have succeeded better with the same problems if mentally mature?

There are many important features of the results which, for lack of space, have not been indicated or discussed. They can be developed from later comparative studies of the data, for in the tables appear all of the essential facts of response apart from those mentioned in the text.

IV

RESULTS OF SUPPLEMENTARY TESTS OF IDEATIONAL BEHAVIOR

1. Julius, Pongo pygmaeus

Box Stacking Experiment In addition to the multiple-choice experiments which have been described in detail in the previous section, it was possible to conduct certain less systematic tests of ideational behavior in the monkeys and the orang utan. From the technical standpoint these tests were relatively unsatisfactory because only inexactly describable. But their results are in many respects more interesting, if not also more important, in the light which they throw on ideation than are those previously presented. First, in order of time, comes a test which may be designated as the box stacking experiment. The method will now be described in connection with an account of the behavior of Julius as contrasted with that of a child of three years and four months of age.

In the large central cage labelled Z, figure 12, which was twenty-four feet long, ten feet wide, and ten to twelve feet deep, the following situation was arranged. From the center of the wire covering of the cage, a banana was suspended on a string so that it was approximately six feet from the floor, five feet from either side of the cage, and twelve feet from either end. From all approaches it was far beyond the reach of Julius, since it was impossible for him to climb along the wire roof and thus reach the string. Two boxes were placed on the floor of the cage several feet from the point directly under the banana. The one of these boxes was heavy and irregular in shape, as is shown in figures 21, 23 and 24 of plate V. Its greatest height was twenty-one inches; its least height, eighteen inches; its other dimensions, twelve and sixteen inches respectively. The smaller and lighter box measured twenty-two by twelve by ten inches. According to the experimenter's calculations, the only way in which Julius could obtain the banana was by placing the smaller box upon the larger and then climbing upon them.

At 10 a.m. on March 5, Julius was admitted to the large cage, and the banana was pointed out to him by the experimenter. He immediately set about trying to get it, and worked diligently during the whole of the period of observation, which, because of the unfinished condition of some of the cages, was limited to slightly over ten minutes. Within this period he made upward of a dozen fairly well directed attempts to obtain the food. Chief among them were three attempts to reach the banana from different positions on the left wall of the cage (as the experimenter faced the laboratory); two attempts to reach it from different positions on the right wall; two from the large box in positions nearly under the banana; two from the large box with the aid of the experimenter's hand; and one from the distant end of the cage(?). There occurred, also, less definite and easily describable efforts to get at the reward.

On account of the unfinished condition of the cages, the experimenter had to remain in the large cage with Julius during the test. This interfered with the experiment because the animal tended both to try to escape and to get the experimenter to help him with his task. Particularly interesting is the latter sort of behavior. After the orang utan had made two or three futile attempts to obtain the food he came to the experimenter, who was standing in one corner of the cage, took him by the hand, and led him to a point directly under the banana. He then looked up toward the banana, grasped the experimenter's arm, raised it, and then tried to pull himself up. He was not allowed to get the food by climbing up on the experimenter. A few minutes later, he again led the experimenter toward the banana, but receiving discouragement in this activity, he proceeded to devote himself to other methods.

Apart from the distractions which have been mentioned above, Julius's attention to the food was surprisingly constant. Whatever his position with respect to it, he seemed not for an instant to lose his motive, and to whatever part of the cage he went and whatever he did during the interval of observation was evidently guided by the strong desire to obtain the banana. Frequently he would look directly at it for a few seconds and then try some new method of reaching it. His gaze was deliberate and in the handling of the boxes he accurately gauged distances. Several times he succeeded in placing the larger box almost directly under the banana, and repeatedly he located that portion of the side wall from which he could most nearly reach the coveted prize.

From my notes I quote the following comment on the results of the initial experiment: "Despite all that has been written concerning the intelligent behavior of the orang utan, I was amazed by Julius's behavior this morning, for it was far more deliberate and apparently reflective as well as more persistently directed toward the goal than I had anticipated. I had looked for sporadic attempts to obtain the banana, with speedy discouragement and such fluctuations of attention as would be exhibited by a child of two to four years. But in less than ten minutes Julius made at least ten obvious and well directed attempts to reach the food. There were also wanderings, efforts to obtain aid from the experimenter, and varied attempts to escape from the cage."

Before proceeding further with the description of the behavior of Julius in the box stacking test, I shall describe for contrast the behavior of a boy three years four months of age when confronted with a situation practically identical with that which the ape was given an opportunity to meet. For the child, the banana was suspended, as previously described, from the roof of the cage. The same two boxes were placed on the floor at considerable distances from the banana, and in addition, a light stick, about six feet long, and a piece of board, the latter by accident, were on the floor. The child was asked to get the banana for Julius, and he eagerly and confidently volunteered to do so.

His behavior may best be described by enumeration of the several attempts made. They include (1) placing the larger box nearly under the banana and reaching from it. (2) Standing of the larger box on end with resulting failure because the child could not stand on the sloping edges of the top of the box. (3) The larger box was turned on its side and the lighter box drawn up opposite it and stood on end. The child then mounted the larger box and from it stepped to the top of the smaller. But the boxes had not been placed beneath the banana, and when the child reached for it, he found himself several feet away from his prize. (4) The boxes were moved to a position nearly under the banana and another futile attempt was made to reach it without placing the smaller box on top of the larger one, the only position from which the child could readily obtain it. (5) The piece of board was placed on top of the larger box and from this height the child again reached upward. (6) The six-foot stick was taken up and an attempt was made to strike the banana and thus dislodge it, but it was too securely fastened to be obtained thus. (7) Attention shifted to other things, and the child played for a time with the board. Reminded of the banana by the experimenter, he again tried method (3). (8) He again used the stick on the banana. (9) The effort to knock the prize to the floor having failed, he became discouraged and said that he must go home. (10) When told that Julius was very hungry and wanted the banana, he repeated efforts similar to those described in (3) and (6).

Up to this time the observations had covered a period of twenty minutes. The child was now taken from the cage and allowed to play about for fifteen minutes. Asked then whether he would go back and try to get the banana, he replied, "No, 'cause I don't want to get it," thus indicating his discouragement with the situation. When taken into the cage, he, nevertheless, made the additional attempts indicated below: (11) Use of one of the boxes. (12) He remarked, "Now I know, I'll get it," and after so saying, repeated (3). (13) Failing, he turned to me and said, "I could get it if I was on your head," but he did not, as Julius had done, lead me to the proper place and try to reach the banana by climbing up or by urging me to lift him. (14) Later, he played in the boxes, apparently forgetful of his task. Finally he remarked: "I'll get the banana," but he made no attempt to do so, and instead, watched the monkeys intently. Thereafter, he showed no further interest in the solution of the problem, and the experiment, after a total period of fifty-five minutes, was discontinued.

Comparison of the behavior of the ape with that of the child indicates a greater variety of ideas for the latter. Julius gauged his distances much more accurately than the child, attended more steadily, and worked more persistently to obtain the reward, but he did not so nearly approach the idea of stacking the boxes as did the child, for the latter, in placing the board on one of the boxes, exhibited in ineffective form the idea which should have yielded the solution of the problem.

The child was given no further opportunity to work at the problem, whereas Julius, as I shall now describe, continued his efforts on subsequent days under somewhat different conditions. On Wednesday, March 10, the banana was suspended as formerly, and three boxes, all of them small and light enough to be readily handled by the ape, were placed in distant parts of the cage. The six-foot stick which had been present in the test with the child, but not in the first test with Julius, was also placed in the cage.

Julius was allowed to work for about an hour. As formerly, he was sufficiently hungry to be eager to get the food and evidently tried all of the possible ways which occurred to him. Chief among these were (1) the use of the various boxes separately or in pairs in very varied positions but never with one upon another,--the only way in which the banana could be reached; (2) climbing to various points on the sides of the cage, with infrequent attempts to reach the banana. Usually his eyes saved him the vain effort.

Unlike the child, Julius paid little attention to the six-foot stick. Two or three times he took it up and seemingly reached for the banana, but in no case did he try persistently to strike it and knock it from the string. It is but fair, however, to remark that such an act is very difficult for the young orang utan, as compared with the child, because of the weakness of the legs and the awkwardness of striking from a sitting posture. As previously, the steadiness of attention and the persistence of effort toward the end in view were most surprising. At one time Julius walked to the end of the cage and there happened to see one of the monkeys eating. He watched intently a few seconds and then hastened back to the banana as if his task had been suggested to him by the sight of the feeding animal. Most interesting and significant in this behavior was the suddenness with which he would turn to a new method. It often looked precisely as though a new idea had come to him, and he was all eagerness to try it out.

On March 11, Julius was given another opportunity to obtain the banana by the use of the three boxes. Although he used them together he made no effort to place one upon another. Certain of his methods are shown in plate V, especially by figures 21, 23 and 24.

This experiment was continued on April 2 under yet different conditions, for this time only two boxes were placed in the cage, the one of them the heavy, irregularly-shaped box and the other the smaller, lighter one originally used. On the end of the heavier box had been nailed a two by two inch wooden block in order to increase the difficulty in using this box alone. As previously, Julius made varied attempts to obtain the banana, but on the whole his interest and attention seemed somewhat weaker than previously and there were indications of discouragement because of repeated failures.

He handled the boxes conspicuously well, and it seemed at times that he would certainly succeed in placing the one upon the other and in reaching the food.

After one series of attempts from the sides of the cage and from the large box, he deliberately turned away from the box and neatly executed a somersault on the floor of the cage, as much as to say, "I am disgusted with the whole situation." Again, later on the same day, after falling from the top of the larger box, which tilted over very easily, he rolled himself into a ball, and childlike, played with his feet. An additional evidence of his changed affective attitude toward his task, especially in connection with definite failures, appeared in his rough handling and biting of the boxes. When most impatient, he worked very roughly.

Julius was allowed to work for the reward from thirty to ninety minutes, or, as a rule, until he had become completely discouraged on April 3, 5, 6, 7, 8, 9 and 13. His behavior was interesting and significant, but nothing new appeared except that his willingness to work gradually disappeared, and on April 13, although previously hungry, he made only a single attempt to obtain the banana and then paid no further attention to it.

The prolonged and varied efforts to obtain the banana were due in a measure at least to three accidental successes. Thus on April 2, 3 and again on the 5th, by fortunate combinations of circumstances, he succeeded in getting the banana, contrary to the intention of the experimenter.

Although active at first on April 6, he soon wearied of his task and quit work. The same was true on April 7, and again on the 8th and 9th. On these days, although hungry, he did not care to enter the large cage and worked only a few minutes each day, seldom making more than two or three half-hearted attempts to obtain the banana. His attitude toward the task had changed completely, in that hopelessness had taken the place of eager expectancy. By the 13th of April he had so nearly given up voluntary efforts to solve the problem that it seemed worth while to test his ability to get the idea by watching the experimenter. For this purpose the following test of imitation was made.

On the morning of April 14, having placed a banana in the usual position, I took Julius into the large cage, dragged the two boxes to the proper position beneath the banana, placed the smaller one upon the larger one and then climbed up on them to show the ape that I could reach the banana. I then stepped down and gave him a chance to climb on the boxes. He did so immediately and obtained the food.

Another piece of banana was supplied, the boxes were placed in distant corners of the cage, and fifteen minutes were allowed Julius so to place them that he could obtain his reward. He gave no indications of having profited by my demonstration, but worked with the boxes singly, usually with the larger one. On April 16, with the banana in position and the two boxes also in the cage, Julius was admitted and allowed to work for five minutes, but again without success. I then placed the boxes properly for him and he immediately climbed up and got the banana. While he was eating, the boxes were carried to distant corners of the cage and another banana placed in position. Now thirty minutes were allowed him for unaided work on the problem. As formerly, the larger box was used repeatedly and attempts to reach from the side of the cage appeared, but there was no tendency to try to use the two boxes together. He worked fairly persistently, however, and showed clearly the stimulating and encouraging effect of aid from the experimenter. Once more, on April 17, Julius was taken into the cage and allowed to watch me place the boxes in proper position. He then climbed up and obtained the desired food. After the bait had been renewed and the boxes displaced, he immediately tried to use the larger one, then he reached for the small one as though to use both together. But the impulse died out and he turned again to the larger box as usual, standing it on end, and persistently trying to balance himself on it. Nothing else of special interest happened during the interval of unaided effort.

Similarly, I placed the boxes for the ape on April 19, allowed him to get the banana and then gave him opportunity to try for himself after the boxes had been displaced. This time he immediately reached for the smaller box and moved it about a little, thus indicating a new association. He next turned to the larger box and worked with it persistently. Later, he once more worked with the smaller box in an unusual manner. He repeatedly stood on it, but made no attempt to lift it or to place it on the larger box. Clearly the usually neglected smaller box had become associated with the satisfaction of obtaining the banana. The same method was carried out on April 20. As I placed the boxes in position beneath the banana, Julius watched with unusual intentness, and when it came his turn to try to obtain the food by the use of the boxes, he began at once to work with the smaller box, but as on April 19, he soon abandoned it and turned to the other. While I was making note of this particular feature of his behavior, he suddenly seized the smaller box by two corners with his hands and by one edge with his teeth, and after a few attempts placed it on top of the larger box, climbed up, and obtained the banana.

Because of bad weather on April 21, the next test was made on April 22, with everything as usual. Unaided, the ape was given an opportunity to obtain the coveted reward, while I stood ready to obtain records of his behavior with my camera. He wasted no time, but piled the smaller box on top of the larger one immediately, and obtained his reward. As soon as opportunity was offered, he repeated the performance. The same thing happened on April 23 and several succeeding dates.

Julius had got the idea, and the only further improvement possible was in skill in manipulating the boxes.

One of the curious performances which appeared during the imitative period is pictured in figure 26, plate V, where the ape is seen lifting the smaller box into the air. This he did three or four times one day, raising it toward the banana each time as though he expected thus to obtain the reward. As he did not go up with the box (according to his expectation?), he abandoned this method, and looking about, discovered the larger box in a distant corner. Thereupon, he promptly pulled the boxes to their proper position beneath the banana, stacked them, and obtained his food.

After considerable skill had been acquired in the placing of the boxes, the one upon the other, the height of the banana above the floor was increased so that three boxes were necessary. Figure 25 of plate V shows him standing on three boxes and reaching upward, and figures 22, 27 and 28 show various modes of handling the boxes and of reaching from them. He was not at all particular as to the stability of his perch, and often mounted the boxes when it seemed to the experimenter inevitable that they should topple over and precipitate him to the floor. Only once, however, during the several days of experimentation did he thus fall.

Obviously important is the evident change in the animal's attention on April 20. He watched with a keenness of interest which betokened a dawning idea.

Before he had succeeded in stacking the boxes, I had written in my note-book, "He seemed much interested today, in my placing of the boxes." Interesting, and important also, is the ease and efficiency with which he met the situation time after time, after this first success. "Trial and error" had no obvious part in the development of the really essential features of the behavior. The ape had the idea and upon it depended for guidance.

Except for the fact that Julius was immature, probably under five years of age, it is likely that he would have stacked the boxes spontaneously instead of by suggestion from the experimenter or imitatively.

No unprejudiced psychologist would be likely to interpret the activities of the orang utan in the box-stacking experiment as other than imaginal or ideational. He went directly, and in the most business-like way from point to point, from method to method, trying in turn and more or less persistently or repeatedly, almost all of the possible ways of obtaining the coveted food. The fact that he did not happen upon the only certain road to success is surprising indeed in view of the many ineffective methods which he used. It seemed almost as though he avoided the easy method.

It is especially important, in connection with these results, to point out the risk of misinterpretation of observations on the anthropoid apes. If they can imitate human activities as readily and effectively as Julius did in this particular experiment, we can never be sure of the spontaneity of their ideational behavior unless we definitely know that they have had no opportunity to see human beings perform similar acts.

Of all the methods of eliciting ideational or allied forms of behavior used in my study of the monkeys and ape, none yielded such illuminating results as the box stacking test, and although from the technical standpoint, it has many shortcomings, as a means to qualitative results it has proved invaluable.

Other Methods of Obtaining the Reward Some weeks later, I tried to discover how Julius would obtain the much desired banana when the boxes were absent. I placed in the large cage a stick about six feet long and an old broom. When admitted, he looked about for the boxes, but not seeing them, picked up the broom and placing it with the splints down, beneath the banana, he tried to climb it, but as it fell over with him, he abandoned this after a few trials, went to his cage, and picking up some old bags which he used at night as covers, he dragged them out and placed them on the floor beneath the banana. He next put the broom upon them and tried to climb up. This general type of behavior persisted for several minutes, everything within reach being used as were the bags, as a means of raising him in the desired direction. Finally, he placed his feet on the broom where the handle joins the splints, seized the handle near the top with his hands, drew himself up as far as possible, and then launched himself in the air and tried to seize the banana. On the third attempt he succeeded.

Later, he was given a plain stick about five feet long. Figure 32 of plate VI shows him using this to obtain the banana in the manner described above. He would grasp it with one or both feet, usually one, ten to fifteen inches from the floor of the cage, meanwhile holding with his hands near the top of the stick. He would then, with all his strength, draw himself up suddenly and jump toward the banana. Often he came down rather hard on the cement floor, much to his disgust.

Yet another method of obtaining the reward developed a day or two later. A light red-wood stick about five feet long and an inch in its other dimensions was the only object in the cage which could possibly be of use in obtaining the banana. The aim of the experimenter was to discover whether Julius would use this as a club.

Previously, in connection with the use of the boxes, he had taken up the same stick two or three times and reached for the banana with it, but in no case had he struck at it or clearly tried to knock it from the string, as did the child most readily and naturally. When provided with this same stick, and it alone, as a means of obtaining the food, he hit upon the following interesting method. Placing one end of the stick between a wooden brace and the wire side of the cage, he climbed up to a level with the banana as is shown in figure 33 of plate VI. Then holding with one hand and one foot to a timber of the cage and to the stick with his other foot, he swung outward as far as possible and reached the banana with his free hand. Having once succeeded by the method, he used it whenever given an opportunity. It was impossible for him to make the reach without the use of the small stick, while with it he succeeded fairly easily and regularly.

Box and Pole Experiment Following the box stacking test, Julius was given an opportunity to exhibit ideation in another type of experiment. This may be designated the box and pole test. The conditions are describable thus. A strong wooden box eighty-four inches long, by four inches wide, by four inches deep, with open ends, was built with one side hinged. Hasps and padlocks enabled the experimenter to lock this "lid" after food had been placed in the center of the box. This box could be placed in the center of the large cage and there fastened by means of cross bars. It is well shown in position in figure 29, plate V. Two poles each eight feet long and approximately one and a half inches in their other dimensions were the only additional materials in the experiment.

On May 1, Julius was allowed to see the experimenter place a half banana in this box, close the lid, lock it in position, and securely fasten the box by means of the cross bars. He was then given opportunity to try to get the banana. The two poles lay on opposite sides of the box and near the edges of the cage. Doctor Hamilton and the writer were in the cage watching. Julius looked into the box through one end, and seeing the banana, reached for it. He could not obtain it in this way, so he began to bite at the box and to pull at it with all his strength. During the fifteen minutes allowed him, he worked at the box in a great variety of ways, fooling with the locks which had been attached to the hasps as well as with the cross bars and continually reaching in at the one or the other end. He was somewhat distracted by the presence of the two observers and attended rather unsatisfactorily to the task in hand. Not once did he touch the poles, and it is doubtful whether he even noticed them. He was not very hungry at this time, and after a few minutes active work he virtually gave up trying to get the food.

Two days later, on May 3, the box was once more placed in position, this time with a half banana in the middle and a small piece of banana near each open end. The two poles lay on the floor of the cage, each several feet distant from the box. Julius was eager for food. When released he went immediately to the box, reached in and obtained a piece of banana from the end nearer the laboratory. He then looked in and saw the piece near the middle of the box. His next move was to pick up the eight foot pole and push it into the box, but before pushing it all the way through, he stopped and began to pull at the box in various ways. Shortly he returned to the pole and twice thrust it in as far as he could reach. The first time, after thrusting it all the way through, he pulled it out and examined the end as though expecting the banana to come out with it. After a third attempt he looked into the box, presumably seeing the banana, then turned a backward somersault, came to the end of the cage, and looked at me. Had it been at all possible, he would have taken me by the hand and led me to the box as a helper. After a few seconds, he returned to the pole, pried the lid of the box with it, then gnawed at the pole. For about five minutes he worked fairly rapidly and steadily, using the poles, pulling, gnawing, and walking about.

His next move was to go to the opposite end of the box, look in, take the piece of banana which was near the opening, then pick up the second pole, which had not previously been noticed, and after a number of attempts, push it into and through the box, looking after it and then pulling it out and looking into the box. Having done this he again came to my end of the cage, and from there returned to try once more with the pole which he had first used. He pushed this pole all the way through, then walked to the other end of the box, looked in and reaching in, obtained the banana which had been pushed far enough along to be within his grasp. Figures 29, 30 and 31 of plate V show stages of this process.

Julius had worked twenty-four minutes with relatively little lost time before succeeding. He had shown almost from the start the idea of using the pole as an instrument, and his sole difficulty was in making the pole serve the desired purpose.

The experiment was rendered still more crucial on May 5 by the placing of the two poles upright in opposite corners of the large cage. For a few minutes after he entered the cage, Julius did not see them, and his time was spent pulling and gnawing at the box. Then he discovered one of the poles, seized it, and pushed it into the box. He tried four times, then went and got the other pole and pushed it into the opposite end of the box. Twice he did this, then he returned to the original pole, bringing the second one with him. He pushed it in beside the first, and as it happened, shoved the banana out of the opposite end of the box. But he did not see this, and only after several seconds when he happened to walk to that end of the box did he discover the banana. The total time until success was fifteen minutes.

Subsequently the ape became very expert in using the pole to obtain the banana, and often only a minute or two sufficed for success. It was not possible for him to direct the stick very accurately, for when he was in such a position that he could look through the box, he could not work the stick itself. It was, therefore, always a matter of chance whether he obtained the banana immediately or only after a number of trials.

Although it is possible that the use of the poles in this experiment was due to observation of human activities, it seems probable in the light of what we know of the natural behavior of the anthropoid apes that Julius would have solved this problem independently of human influence. It was the expectation of the experimenter that the pole would be used to push the banana through the box, but as a matter of fact the ape used it, first of all, to pull the food toward him, thus indicating a natural tendency which is important in connection with the statements just made. Subsequently he learned that the banana must be pushed through and obtained at the farther end of the box. I am not prepared to accept the solution of this problem as satisfactory evidence of ideation, but I do know that few observers could have watched the behavior of the orang utan without being convinced that he was acting ideationally.

Draw-in Experiment

An interesting contrast with the box and pole test is furnished by what may be called the draw-in experiment. This was planned as a simple test of Julius's ability to use a stick to draw things into his cage from beyond the wire side. A board was placed, as is shown in figure 34 of plate VI, with sides to hold a banana, carrot, or some other bit of food, in position. In the actual test either a carrot or a banana was placed about two feet from the wire netting and a stick two feet long was then put into the cage with the ape.

When this situation was first presented to Julius, he looked at the banana, reached for it, and failing, picked up a bag from the floor of the cage and tried to push it through the wire mesh toward the banana. He also used a bit of wire in the same way, but was unable thus to get the food. As soon as a stick was placed in his cage, he grasped it and used it in a very definite, although unskillful, way to pull the banana toward him. He was extremely eager and impatient, but nevertheless persistent in his efforts, and within five minutes from the beginning of the first trial, he had succeeded in getting two pieces of banana, using always his left hand to manipulate the stick. This test was repeated a number of times with similar results. He had from the first the ability to use a stick in this way, and the only difficulty with the test as a means of obtaining evidence of ideational behavior is that the possibility of imitation of man cannot be certainly excluded.

Lock and Key Test By my assistant it was reported on May 5 that the orang utan had been seen to place a splinter of wood in a padlock which was used on the cages and to work with it persistently. It looked very much like imitation of the human act of using the key, and I therefore planned a test to ascertain whether Julius could readily and skillfully use a key or could learn quickly to do so by watching me.

The first test was made on May 15 with a heavy box whose hinged lid was held securely in position by means of a hasp and a padlock. The key, which was not more than an inch in length, was fastened to a six inch piece of wire so that Julius could not readily lose it. With the animal opposite me, I placed a piece of banana in the box, then closed the lid and snapped the padlock. I next handed Julius the key. He immediately laid it on the floor opposite him and began biting the box, rolling it around, and occasionally biting also at the lock and pulling at it. During these activities he had pulled the box toward his cage. Now he suddenly looked up to the position where the banana had been suspended in the box experiment. Evidently the box had suggested to him the banana. For thirty minutes he struggled with the box almost continuously, chewing persistently at the hinges, the hasp, or the lock. Then he took the key in his teeth and tried to push it into one of the hinges, then into the crack beneath the lid of the box.

Subsequently I allowed him to see me use the key repeatedly, and as a result, he came to use it himself now and then on the edge of the box, but he never succeeded in placing it in the lock, and the outcome of the experiment was total failure on the part of the animal to unfasten the lock of his own initiative or to learn to use the key by watching me do so. I did not make any special attempt to teach him to use the key, but merely gave him opportunity to imitate, and it is by no means impossible that he would have succeeded had the key been larger and had the situation required less accurately coordinated movements. However, it is fair to say that the evidence of the idea of using the key in the lock was unconvincing. My assistant's observation was, perhaps, misleading in so far as it suggested that idea. It may and probably was purely by accident that the animal used the splinter on the padlock.

2. Skirrl, Pithecus irus

Box Stacking Experiment The monkey Skirrl was tested by means of the box stacking experiment much as Julius had been. On August 23, with a carrot suspended six feet from the floor of the large cage and three boxes in distant corners, the animal was admitted and his behavior noted.

The boxes, which were made of light, thin material, ranged in size from one six inches in its several dimensions to one twenty inches long, thirteen inches wide, and eleven inches deep. Only by using at least two of these boxes was it possible for the animal to reach the carrot.

Immediately on admission to the cage, Skirrl began to gnaw at the boxes, trying with all his might to tear them to pieces. After some thirty minutes of such effort, interrupted by wanderings about the cage and attempts to get at the other monkeys, he suddenly went to the largest box of all, set it up on end almost directly under the carrot, mounted it, and looked up at the food. It was still beyond his reach and he made no effort to get it, but instead, he reached from his perch on the big box for the next smaller box, which was approximately sixteen inches, by fourteen, by twelve. This he succeeded in pulling toward him, at the same time raising it slightly from the floor, but his efforts caused the large box to topple over and he quit work. The experiment was discontinued after a few minutes, the total period of observation having been thirty-five minutes.

Skirrl handled the boxes with ease and with evident pleasure and interest. He also noticed the carrot at various times during the interval, but his attention was fixed on it only for short periods.

The test was continued on August 24 when, instead of a carrot, a half banana was used as bait. It was placed only five feet from the floor, and three boxes were as formerly placed in distant corners of the cage. When admitted, Skirrl looked at the banana, then pulled one of the boxes toward it, but instead of mounting, he went to the smallest box and began to gnaw it.

Shortly, he mounted the middle sized box and looked up toward the banana, but the box was not directly under the bait, and in any event, it would have been impossible for him to reach it. He next went to the largest box, gnawed it vigorously, turned it over several times, and then abandoned it for the middle sized box, from which by skillful use of his teeth and hands, he quickly tore off one side.

By this time, apparently without very definitely directed effort on the part of the monkey, all three of the boxes were in the center of the cage and almost directly beneath the banana. Skirrl climbed up on the largest box and made efforts to pull the middle sized one up on to it, the while looking at the banana every few seconds. He did not succeed in getting the boxes properly placed, and after a time began moving them about restlessly.

His behavior plainly indicated that hunger was not his chief motive. He was more interested in playing with things or in working with them than in eating, and the satisfaction of tearing a box to pieces seemed even greater than that of food. It is especially noteworthy that when Skirrl attempts to dismember a box, instead of starting at random, he searches carefully for a favorable starting point, a place where a board is slightly loosened or where a slight crack or hole enables him to insert his hand or use his teeth effectively. Many times during this experiment he was observed to examine the boxes on all sides in search of some weak point. If no such weak point were found, he shortly left the box; but if he did find a favorable spot, he usually succeeded, before he gave up the attempt, in doing considerable damage to the box.

Following the behavior described above, Skirrl returned to the middle sized box, placed it on end under the banana, mounted, and looked upward at the bait, but as it was a few inches beyond his reach, he made no attempt to get it, but instead, after a few seconds, went to the smallest box, and finding a weak point, began to tear it to pieces.

Later he rolled what was left of the smallest box close to the other two boxes, nearly under the banana, and the remainder of his time was spent gnawing at the boxes and playing with pieces which he had succeeded in tearing from them. During the remainder of the thirty minute interval of observation, no further attention was given the bait.

Again, on August 25, the test was tried, but this time with boxes whose edges had been bound with tin so that it was impossible for the monkey to destroy them. He spent several minutes searching for a starting point on the middle sized box, but finding none, he dragged it under the banana, looked up, mounted the box, but, as previously, did not reach for the bait because it was beyond his reach. He then played with the boxes for several minutes. Finally he worked the two smaller boxes to a position directly under the banana, put the middle sized one on end, mounted it, and looked at the bait, but again abandoned the attempt without reaching.

During the thirty minutes of observation he made no definite effort to place one box upon another. Three times he mounted one or another of the boxes when it was under the banana or nearly so, but in no case was it possible for him to reach the bait.

From the above description of this monkey's behavior, it seems fairly certain that with sufficient opportunity, under strong hunger, he would ultimately succeed in obtaining the bait by the use of two or more boxes. For his somewhat abortive and never long continued efforts to drag two boxes together or to place the one upon the other clearly enough indicate a tendency which would ultimately yield success. The possibility of imitation is not excluded, for Skirrl had opportunities to see Julius and the experimenter handle the boxes.

Because of the other work which seemed more important at the time, this experiment was not continued further. The results obtained suggest the desirability of testing thoroughly the ability of monkeys to use objects as only the anthropoid apes and man have heretofore been thought capable of using them.

Box and Pole Experiment Skirrl was first tested with the box and pole experiment on August 12. As in the case of Julius, a half banana was placed in the middle of the long box and the attention of the monkey was attracted to the bait by small pieces of carrot placed near each open end. Two poles were placed near the box on the floor of the cage. When admitted to the cage Skirrl went almost directly to the ends of the box, took the pieces of carrot which were in sight, but apparently failed to perceive the bait in the middle of the box. For a while he played with the locks on the box, shoved it about, and amused himself with it, showing no interest in obtaining the food. Later he looked through the box and saw the banana. He then dragged the box about, apparently trying to get it into his cage, but he gave no attention to the poles nor did he make any evident effort to obtain the banana which was easily visible in the center of the box. The period of observation was only twelve minutes.

On August 24 this experiment was repeated with an important modification of the apparatus in that the wooden lid of the long box had been replaced by a wire cover through which the animal could see the bait. Two poles were as formerly on the floor of the cage, not far from the box. Skirrl almost immediately noticed the banana and tried to get it by gnawing at the box. He did not once reach in at the ends of the box, but he did handle the poles, throwing them about and pounding with them. There was not the slightest attempt to use them in obtaining the bait.

This experiment was later repeated three times at intervals of a number of days, but in no case did Skirrl show any tendency to use the poles as means of obtaining the food.

Draw-in Experiment

This also was arranged in the same manner as for Julius, and on each of five days Skirrl was allowed at least thirty minutes to work for the bait. Either a banana or a carrot was each day placed on the board well beyond his reach, and one or two, usually two, small sticks were put into his cage. Not once during the several periods of observation did Skirrl make any attempt to use a stick or any other object as a means of drawing the food to him. Instead, he reached persistently with his arm, pulled and gnawed at the wires which were in his way, and occasionally picked up and gnawed or pounded with the sticks in the cage. His attention every now and then would come back to the food, but it tended to fluctuate rather rapidly, and in the regular period of observation, thirty minutes, it is unlikely that he attended to the bait itself for as much as five minutes. In this respect as well as many others, Skirrl's behavior contrasts sharply with that of the orang utan.

The results of this experiment indicate the lack in the monkey of any tendency or ability, apart from training, to use objects as means of obtaining food. Ways of using objects as tools which apparently are perfectly natural to the anthropoid apes and to man are rarely employed by the lower primates.

Hammer and Nail Test One day I happened to observe Skirrl playing with a staple in his cage. He had found it on the floor where it had fallen and was intently prodding himself with the sharp points, apparently enjoying the unusual sensations which he got from sticking the staple into the skin in various portions of his body, and especially into the prepuce.

A few days later I saw him playing in similar fashion with a nail which he had found, and still later he was seen to be using a stick to pound the nail with. This suggested to me the hammer and nail test.

A heavy spike was driven into an old hammer to serve as an indestructible handle. This hammer, along with a number of large wire nails and a piece of redwood board, was then placed in the monkey's cage. Skirrl immediately took up the hammer, grasping the middle of the handle with his left hand, and with his right hand taking up a nail. He then sat down on the board, examined the nail, placed the pointed end on the board, and with well directed strokes by the use of the head of the hammer drove the nail into the board for the distance of at least an inch. He then tried to pull it out, but was forced to knock it several times with the hammer before he could do so.

This performance, during the next few minutes, was repeated several times with variations. Often the side of the hammer was used instead of the head, and occasionally, as is shown in figure 8 of plate II, he seized the hammer well up toward the juncture of the same with the spike. This figure does justice to the performance. At the moment the picture was taken, Skirrl's attention had been attracted by a monkey in an adjoining cage, and he had momentarily looked up from his task, the while holding nail and hammer perfectly still.

This test was repeated on various days, and almost uniformly Skirrl showed intense interest in hammer and nails and used them more or less persistently in the manner described. Occasionally, apparently for the sake of variety, he would put the blunt end of the nail on the board and hammer on the point. Again, he would try persistently to drive the nail into the cement floor, and once by accident, when hammer and nails were left in his cage over night, he succeeded in making several holes in the bottom of his sheet iron water pan.

There was no doubting the keen satisfaction which the animal took in this form of activity.

It is impossible to say that the behavior was not imitative of man, for Skirrl, along with all of the other monkeys, had had abundant opportunity to see carpenters working. But this much can be said against the idea of imitation,-- no one of the other animals, not excepting the orang utan, showed any interest whatever in hammer and nails. Occasionally they would be played with momentarily or pushed about, but Sobke, Jimmie, Gertie, Julius, although given several opportunities to exhibit any ability which they might have to drive nails, made not the least attempt to do so. Evidently we must either conclude that Skirrl had a peculiarly strong imitative tendency in this direction, or instead, a pronounced disposition or instinct for the use of objects as tools. It would seem fair to speak of it as an instinct for mechanical activity.

Under this same heading may be described Skirrl's reactions to such objects as a handsaw, a padlock, and a water faucet. The saw was given to him in order to test his ability to use it in human fashion, for if he could so expertly imitate the carpenter driving nails, it seems likely that he might also imitate the use of the saw.

As a matter of fact, he showed no tendency to use the saw as we do. Instead, he persistently played with it in various ways, at first using it as a sort of plane to scrape with, later often rubbing the teeth over a board so that they cut fairly well, but never as effectively as in the hands of a man. After two or three days' practice with the saw, Skirrl hit upon a method which is, as I understand, used by man in certain countries, namely, that of placing the saw with the teeth up, holding it rigid, and then rubbing the object which is to be sawed over it. This Skirrl succeeded in doing very skillfully, for he would sit down on the floor of the cage, grip with both feet the handle of the saw, with the teeth directed upward, then holding either end in his hands, he would repeatedly rub a stick over the teeth. In this way, of course, he could make the saw cut fairly well. But still more to his liking was the use of a spike instead of a stick as an object to rub over the teeth, for with this he was able to make a noise that would have satisfied even a small boy.

Further light is shed on the force of the tendency to imitate man by the saw test. After Skirrl had been given an opportunity to show what he could do with the tool spontaneously, I demonstrated to him the approved human way of sawing. Often he would watch my performance intently as though fascinated by the sound and motion, but when given the tool he invariably followed his own methods. Although I repeated this test of imitation several times on three different days, the results were wholly negative.

Other Activities One day as Skirrl was being returned to his own cage by way of the larger cage, he picked up an unfastened padlock and carried it into the cage with him. For more than an hour he amused himself almost without interruption by playing with this lock. The things which he did with it during that time would require pages to describe. His interest in it was very similar to that which he had exhibited in hammer and nails, saw, and indeed any objects which he could play with. The lock was pounded in various ways, bitten, poked with nails, hooked into the wires of the cage, used to pull on, pounded with a stick, used to hammer on the floor of the cage with, and in fine, manipulated in quite as great a variety of ways as a human being could have discovered. Finally it was hooked to the side of the cage and snapped shut, and as Skirrl was unable to dislodge it from this position, he shortly gave up playing with it.

At the end of the large cage and just outside the wire netting was a faucet to which a hose was usually attached. The valve could be opened by turning a wheel-shaped hand piece. Both Skirrl and Julius learned to turn this wheel in order to get water to play with, but usually the former's strength was not sufficient to turn on the water. The latter could do it readily. The indications are that both animals profited by seeing human beings turn on the water. This unquestionably attracted their attention to the faucet, and probably by playing with it they accidentally happened upon the proper movement. At any rate, Skirrl's behavior was significant in this connection, for he would pick up the hose to see if water were flowing, and if it were not, he would throw it down, go directly to the faucet, and try to turn the wheel. The association of the wheel with the desired flow of water was therefore definitely established. Shall we describe the act as ideational? It seems the natural thing to do.

3. Sobke, Pithecus rhesus

Box Stacking Experiment For this test, in the case of Sobke, three light boxes made of redwood about one-third of an inch thick were used. The smallest, box 1, was six inches in each direction, the next larger, box 2, was twelve inches, and the third, box 3, eighteen inches. As in the case of the other animals, bait, either banana or carrot, was suspended from the middle of the roof of the large cage at such distance from the floor as to be reached by the animal only by the use of the boxes.

The first observations on Sobke were made on June 14. The three boxes had been placed in the form of a pyramid directly under the banana, which hung about eighteen inches above the uppermost box. Sobke's attention while in his cage had been attracted to the bait by seeing me fastening it in position, but when admitted to the large cage, he simply glanced at it and then wandered about the cage, picking up bits of food and struggling to get at the other monkeys. This he did for about five minutes. He then went to the boxes, placed his hands on top of the bottom one, but did not climb up on it. A few minutes later he returned to the box again, climbed up, and readily reached the food, which he ate while resting on boxes 1 and 2.

I now replaced the bait and gave the monkey a second chance to obtain it. Almost immediately he climbed up as far as the second box, but although he could reach the banana only from the uppermost box, he deliberately shoved it off to the ground and sat down upon box 2. As he was unable to obtain the banana from this, he soon began to gnaw and pull at it, and as he was succeeding all too well in his efforts to tear the box to pieces, he had to be returned to his cage.

The most important features of his behavior were, first, his stealthy and indirect manner, and second, his failure to use other means of obtaining the bait than that supplied by the observer. Instead of looking straight at the experimenter, or at the object which he wished to obtain, he apparently looked and attended elsewhere. For this reason it was often difficult to decide whether or not he had noticed the bait or the boxes. Finally I was led to conclude that he usually knew exactly what was going on and had in his furtive way noted all of the essential features of the situation, and that his manner was extremely indicative of his mental attitude of limited trust. Both Julius and Skirrl went to the opposite extreme in the matter of directness, or as we should say in human relations, frankness. They would look the experimenter directly in the eye, and they usually gazed intently at anything, such for example as the bait, that interested them. Sobke, even when very hungry, instead of going directly toward the bait, and trying to obtain it, usually did various other things as though pretending that he had no interest in food.

On the following day, June 15, the three boxes were again placed nearly under the banana, but this time the two smaller boxes, numbers 1 and 2, were pushed to the extreme end of the lower box and so far from the bait that it could not be reached from box 1. It was necessary then for the animal to push boxes 1 and 2 along on box 3 until they were nearer the bait.

Sobke, when admitted to the cage, evidently noticed the banana, but as formerly, he made no immediate effort to obtain it. After wandering in search of food and quarreling with the other monkeys for several minutes, he went to the boxes, pushed the topmost one, number 1, off on to the floor, and then carried it into his cage where he quickly tore one side off. He next returned to the large cage, climbed up on box 2, and he was able, by jumping, to reach and obtain the banana.

As Sobke was very good at jumping, his new method rendered the box stacking experiment of uncertain value, since it was next to impossible so to arrange the spatial relations of bait and boxes that he should be neither discouraged by too great a distance nor encouraged to jump by too small a distance. Evidently it would be more satisfactory to simplify the conditions by trying to discover, first of all, whether he would use a single box as a means of reaching the reward.

In pursuance of this idea, I suspended a piece of bread five feet from the floor of the cage, and a few feet to one side of it, I placed a box from which it could be reached, or at least easily seized by jumping. Sobke shortly walked to a point beneath the bait and leaping into the air, seized it.

I then replaced the bait, raising it to a height of five feet ten inches from the floor of the cage. When I had retired, Sobke placed himself in the proper position beneath, looked up at it, but went away without jumping for it. During the remaining ten minutes of observation, he paid no further attention to the bait, having satisfied himself evidently that it was beyond his reach.

My use of this test was concluded on June 16 when once more I suspended a piece of bread six feet from the floor and placed a few feet to one side the eighteen inch box, number 3, from which had the monkey pushed it to a point directly under the bread, he could have obtained the food easily. Sobke noticed the food promptly, and from time to time as he wandered about, he glanced at it out of the corner of his eye, but not once did he sit down and look at it steadily and directly as Julius and Skirrl might have done.

In the first twenty minutes of observation the monkey made no attempt either to use the box or to reach the food by jumping. I then placed the box directly under the bait, and scarcely had I withdrawn from the cage before Sobke climbed up on it and looked toward the food. He could not reach it without jumping, and he made no effort to get it. I had left a second box in the cage,--one which I had been using as a seat. Sobke now went to this box, placed his hands on it, looked toward the bait, and then went to a distant part of the cage. No further indications were obtained during the remainder of the period of observation of interest in the boxes as possible means of obtaining the desired food.

It is of course obvious that this experiment was not long enough continued to justify the conclusion that either Sobke or Skirrl could not use the boxes or even learn to place one box upon another in order to obtain the bait. The experiment, like several others which are being described briefly, was used to supplement the multiple-choice experiment, and the experimenter's chief interest was to discover the number and variety of methods which would be used by the animal in the first few presentations of a situation. It is practically certain that both of these monkeys would have succeeded ultimately in solving the problem of obtaining the food had they been left in the cage with a number of boxes, for Skirrl very early indicated interest in moving the boxes about, and Sobke showed a tendency in that direction which perhaps was inhibited partially by his distrust of the experimenter.

Draw-in Experiment

For Sobke, as for Julius and Skirrl, the draw-in test was made by putting food on a shelf outside the cage, beyond the reach of the animal, and placing in the cage with the animal one or two sticks long enough to be used for drawing in the bait.

Sobke was first given this test on July 24. He tried persistently to reach the banana with his hand, seized the box which supported the bait, shook it, picked up one or other of the sticks, and chewed at it repeatedly, but not once did he make any move to use a stick to draw the food toward him.

This experiment was repeated on July 27, 29, 30 and 31, a period of thirty minutes being allowed on each day for observation. At no time did Sobke show any inclination to use either a stick or any other object as a means of reaching the bait. Instead, he confined himself strictly to the use of hands and teeth.

This test makes it fairly certain that Sobke had no natural tendency to use objects as tools. In so far as he attended to things about the cage or laboratory, it seemed to be rather to play with them in a general way than to use them ideationally or otherwise for definite purposes.

The definitely negative result of the draw-in experiment rendered needless prolonged observation with the box and pole test, whose results are now to be presented.

Box and Pole Experiment The eighty-four inch box, previously used for a similar test with Julius, was presented to Sobke on August 24, the wooden cover having been replaced by a wire one so that the monkey could readily see the bait in the middle of the box. Sobke, when admitted to the large cage, went directly to the box and at once discovered the banana which was midway between the ends. He evidently desired it. Shortly, he went to one end of the box and looked in. This he repeated later. He also shook the box and tried to pull it about and tear it with his teeth, but to the two poles lying nearby on the floor of the cage he gave not the slightest attention during a thirty minute period of observation.

The experiment was not repeated because of more important work.

Other Activities In more respects than I have taken time to enumerate in the above descriptions of behavior, the relations of Sobke to objects differed from those of Skirrl, and still more from those of Julius. Hammer, nails, saw, stones, sticks, locks, and various other objects received relatively little attention from Sobke unless they happened to come in his way; then they were usually pushed aside with but scant notice. Rarely he would carry something to the shelf of his cage with him, but as a rule only to lay it down and attend to something else. Skirrl, on the contrary, attended persistently to anything new in the shape of a movable object. He was extremely partial to objects which could be manipulated by him in various ways, and especially to any thing with which he could make a noise. His interest in hammer and nails, saw, locks, etc., seemed never to wane. I have seen him play for an hour almost uninterruptedly with a hammer and a nail, or even with a big spike which he could use to pry about his cage. In the absence of anything more interesting, even a staple or a small nail might receive his undivided attention for minutes at a time. How important is the species difference in this connection, I have no means to judge, but if we may not consider these different modes of behavior characteristic of _P. rhesus_ as contrasted with _P. irus_, we must conclude that remarkable individual differences exist among monkeys, for whereas Skirrl is by nature a mechanical genius, Sobke has apparently no such disposition. I can imagine no more fascinating task than the careful analytical study of the temperaments of these two animals. Skirrl's behavior has importantly modified my conception of genius.

V

MISCELLANEOUS OBSERVATIONS

1. _Right- and left-handedness_

Several years ago Doctor Hamilton reported to me observations which he had made on preference for the right or left paw in dogs. He has not, I believe, published an account of his work. Subsequently, Franz observed a similar preference in monkeys which, according to his report, exhibit marked tendency to be right-handed, left-handed, or ambidextrous.

My own observations, although they are wholly incidental to my other work, seem worthy of description at this point. I noted, first of all, that the orang utan Julius tended to use his left hand. He by no means limited himself to this, but in difficult situations he almost invariably reached for food or manipulated objects in connection with food getting with the left hand.

Figures 23 and 24 of plate V, show him reaching for a banana with the left hand. Likewise, figure 34 exhibits the use of the left hand in the draw-in experiment.

So marked was Julius's preference for his left hand that I became interested in observing similar phenomena in the monkeys. Skirrl, when driving nails, held the hammer with his left hand and the nail with his right hand. The fact that he never was observed to reverse the use of the hands is surprising, for other observations indicate that he preferred the right hand for certain acts.

Stimulated by the obvious left-handedness, in certain connections, of Julius and Skirrl, I tested the preference of several of the monkeys in the following simple way. Standing outside the cage I would hold out a peanut to a hungry animal, keeping it so far from the cage that the monkey could barely reach it with its fingers. I noted the hand which was used to grasp the food. Next I varied the procedure by placing the peanut on a board in order to make sure that I was not definitely directing the animal's attention.

With Sobke the following results were obtained. In forty trials given on two different days, he reached for and obtained the food each time with his left hand. Only by holding the bait well toward the right side of his body was it possible to induce him to use the right hand. So far as may be judged from these observations and from others in connection with the experiments, this animal is definitely left-handed.

With Skirrl the results are strikingly different. As stated above, he used the hammer consistently with his left hand, but in twenty attempts to obtain food by reaching, he used his right hand seventeen times and his left only three times. It was quite as difficult to induce him to use his left hand for this purpose as it was to induce Sobke to use his right. We must therefore conclude that Skirrl is right-handed in connection with certain movements and left-handed in others.

The monkey named Gertie in the reaching experiment consistently used her left hand, never once using the right.

Jimmie, so far as it was possible to make tests with him, also used his left hand, but it should be said that the results are unsatisfactory because he was at the time extremely pugnacious and paid attention to the experimenter rather than to the food.

Scotty, in the first series of ten trials, used his right hand eight times, his left twice. In the second series, given the following day, he used the right hand three times and the left seven times. From this we should have to infer that he is ambidextrous.

A female rhesus monkey which had been brought to the laboratory only a few days previously showed a preference for the right hand by the use of it fourteen times to six.

In connection with these data which are, I should repeat, too scanty to be of any considerable value, I wish to describe my own experience. Although naturally left-handed, I am by training right-handed to the extent of having been able to use my hands in writing and in various other activities equally well at the age of twelve. I am at present ambidextrous in that there are many things which I do with equal readiness and skill with either hand. Delicate, exact, and finely coordinated movements, such as those of writing and using surgical instruments, I perform always with my left hand while grosser movements involving the whole hand or arm, I am rather likely to perform with my right hand.

It seems not improbable in the light of my own experience that we shall find some specialization among the lower animals with respect to preference for right and left hand or arm. I should not be at all surprised to discover that it is the rule for animals to possess or to develop readily definite preference for one hand in connection with a given act of skill and to have quite as definite a preference for the other hand in connection with a radically different kind of act.

2. Instinct and emotion Of the many presumably instinctive modes of behavior which were observed, only those which have to do with social relations seem especially worth reporting. From among them I shall select for description a few which have already been referred to in connection with the experimental observations.

Maternal Instinct Aspects of the maternal instinct I had opportunity to observe in Gertie, who on February 27 gave birth to a male infant, I present below the substance of a previously published note on her behavior (Yerkes, 1915).

"On February 27 one of the monkeys of our collection gave birth, in the cages at Montecito, to a male infant. The mother is a Macacus cynomolgus rhesus (_P. irus rhesus_) who has been described by Hamilton (1914, p. 298) as 'Monkey 9, Gertie, _M. cynomolgus rhesus_ (_P. irus rhesus_). Age, 3 years 2 months. (She is now, May 1, 1915, 4 years and 6 months.) Daughter of monkeys 3 and 10. First pregnancy began September, 1913.' The result of this pregnancy was, I am informed, a still-birth.

"The second pregnancy, which shall now especially concern us, resulted likewise in a still-birth. Parturition occurred Saturday night, and the writer first observed the behavior of the mother the following Monday morning. In the meantime the laboratory attendant had obtained the data upon which I base the above statements.

"At the time of parturition Gertie was in a 6 by 6 by 12 foot out-door cage containing a small shelter box, with an exceptionally quiet and gentle male (not the father of the infant) who is designated in Hamilton's paper as Monkey 28, Scotty.

"My notes record the following exceptionally interesting and genetically important behavior. On March 1, when I approached her cage, Gertie was sitting on the floor with the infant held in one hand while she fingered its eyelids and eyes with the other. Scotty sat close beside her watching intently. When disturbed by me the mother carried her infant to a shelf at the top of the cage. Repeatedly attempts were made to remove the dead baby, but they were futile because Gertie either held it in her hands or sat close beside it ready to seize it at the slightest disturbance.

"Especially noteworthy on this, the second day after the birth of the infant, are the male's, as well as the female's, keen interest in the body and their frequent examinations of the eyes, as if in attempts to open them. Often, also, the mother searched the body for fleas.

"Observations were made from day to day, and each day opportunity was sought to remove the body without seriously frightening or exciting the female. No such opportunity came, and during the second week the corpse so far decomposed that, with constant handling and licking by the adults, it rapidly wore away. By the third week there remained only the shriveled skin covering a few fragments of bone, and the open skull from the cavity of which the brain had been removed. This the mother never lost sight of: even when eating she either held it in one hand or foot, or laid it beside her within easy reach.

"Gradually this remnant became still further reduced until on March 31 there existed only a strip of dry skin about four inches long with a tail-like appendage of nearly the same length.

"The male, Scotty, on this date was removed to another cage. Gertie made a great fuss, jumping about excitedly and uttering plaintive cries when she discovered that her mate was gone. Whenever I approached her cage she scurried into the shelter box and stayed there while I was near. This behavior I never before had observed. It continued for two days. On April 2, it was noted that she had lost her recently acquired shyness and she no longer made any attempts to avoid me. As usual, on this date, she was carrying the remnant about with her.

"The following day, April 3, Gertie was lured from her cage to a large adjoining compartment for certain experimental observations. After she had been returned to her own cage the remnant was noticed on the floor of the large cage. I picked it up. Gertie evidently noticed my act; for although at a distance of at least ten feet from me, she made a sharp outcry and sprang to the side of the cage nearest me. I held the piece of skin (it looked more like a bit of rat skin than the remains of a monkey) out to her and she immediately seized it and rushed with it to the shelf at the top of the cage.

"Two days later the remnant was missing, and careful search failed to discover it in the cage. It is probable that Gertie had carelessly left it lying on the floor whence it was washed out when the cages were cleaned. On this date Gertie seemed quieter than for weeks previously.

"Thus it appears that during a period of five weeks the instinct to protect her offspring impelled this monkey to carry its gradually vanishing remains about with her and to watch over them so assiduously that it was utterly impossible to take them from her except by force.

"After reading this note in manuscript, Doctor Hamilton informed me that Gertie had behaved toward her first still-birth as toward her second. And, further, that Grace, a baboon, also carried a still-birth about for weeks.

"I am now heartily glad that my early efforts to remove the corpse were futile, for this record of the persistence of maternal behavior seems to me of very unusual interest to the genetic psychologist."

Fear In connection with the multiple-choice experiments Skirrl exhibited what seemed to be instinctive fear as a result of his unfortunate experience with nails in the floor of box 1. He seemingly referred his misadventure to some unseen enemy under the floor, and this in spite of the fact that he was given abundant opportunity to examine the floor of the box, but not until after the dangerous nails had been clinched. His long continued avoidance of the experiment boxes and his still more persistent hesitancy in entering them, coupled with his almost ludicrous efforts to see beneath the floor through the holes cut for the staples on the doors, gave me the impression of superstitious fear of the unseen. As I watched and recorded his behavior day after day during the period of most pronounced fear, I could not avoid the thought that the instinctive fear of snakes had something to do with his peculiar actions, and although I have never studied either the natural or the acquired responses of monkeys to snakes, I suspect that lacking such instinctive equipment, Skirrl would have behaved differently as a result of the pricks which he received from the nails. It is needless to redescribe his acquired fear of whiteness as it manifested itself in the freshly painted apparatus. Accompanying these instructive modes of response and their emotions are suggestions of peculiarly interesting problems as well as of modes of attacking them. As a matter of fact, Skirrl's fear-reactions did much to alter my conception of the constitution of his mind. I should not have been surprised by the features of behavior exhibited, but I was by no means prepared for their persistence, and for the highly emotional attitude toward the particular situation. Only an organism of complexly constituted nervous system and fairly highly developed affective life could be expected to respond as did this monkey. As has been suggested above, I find the appeal to instinct, modified by experience, a natural mode of accounting for the unexpected features of Skirrl's behavior.

Sympathy The instinctive playfulness of the young monkey Tiny contrasted most strikingly with the more serious, if not more sedate, modes of behavior of the older individuals.

During the greater part of my period of observation Tiny was cage-mate of Scotty, the most calm and apparently lazy of all the monkeys. Tiny delighted in teasing Scotty, and her varied modes of mildly tormenting him and of stirring him to pursuit or to retaliation were as interesting as they were amusing. Her most common trick was to steal up behind him and pull the hair of his back, or seize his tail with her hands or teeth. Often when he was asleep she would suddenly run to him, give a sudden jerk at a handful of hairs, and leap away. He was surprisingly patient, and I never saw him treat her roughly in retaliation.

Another of Tiny's favorite forms of amusement was that of trying to stir up the other monkeys to attacks on one another. She very cleverly did this by pretending that she herself was being attacked. The instant the older animals began to show hostility toward one another she would leap out of the way and watch the disturbance with evident satisfaction. It was this mode of behavior in the little animal which ultimately provided opportunity for the observations which I wish now to report as indicative of sympathetic, possibly I may say altruistic, emotions.

Tiny was confined with Scotty in a cage adjoining the one in which Jimmie and Gertie were being kept. The cages were separated by wire netting of half-inch mesh.

One morning as I was watching the behavior of the animals in the several cages, I noticed Tiny dressing with her teeth a wounded finger. It had evidently been bitten by one of the other animals, in all probability either by Jimmie or Gertie. Tiny was trimming away the loose bits of skin very neatly and cleansing the wound. After working at this task for a few minutes, she quickly climbed up to the shelf near the top of her cage, and by rushing to the partition wire between the two cages, she lured Gertie to an attempted attack on her. Gertie sprang up to the partition, placed her hands on it, with the fingers projecting through the meshes, and attempted to seize Tiny's fingers with her teeth. But the latter was too quick for her, and withdrawing her hands, like a flash seized in her teeth the middle finger of Gertie's left hand. She then bit it severely and with all her might, at the same time pulling and twisting violently, often placing the entire weight of her body on the finger. Her sharp teeth cut to the bone, and it was impossible for the larger and stronger monkey to tear away. For several seconds this continued, then Gertie succeeded in escaping, whereupon she at once retreated to the opposite end of her shelf and proceeded to attend to her injured finger. She cried, wrung her hands, and from time to time placed the finger in her mouth as though in an effort to relieve the pain. By this time Jimmie's attention had been attracted by the disturbance and he rushed up to the shelf, and facing Gertie, watched her intently for a few seconds. The look of puzzled concern on his face was most amusing. Apparently he felt dimly that something in which he should have intelligent interest was going on, but was unable wholly to understand the situation. After watching Gertie for a time and trying to discover what she was doing, which was rendered difficult by her tendency to turn away from him, in order to shield her injured finger, he rushed over to the wire partition and made strenuous efforts to seize Tiny with his hands and teeth. But although she continued close to the partition and often crowded against it with face and hands flattened on the wires, he was not able to get hold of her, and after a few vain attempts he returned to his mate, and again with evident solicitousness and the most troubled expression, watched her wringing her hands and chewing or sucking at her injured finger. Shortly he again returned to the partition and renewed his attempts to seize the young monkey. Thus he went back and forth from one place of interest to the other several times, but being unable to achieve anything at either point, he finally gave up and returned to his breakfast on the floor of the cage.

I report this incident fully because the behavior of Jimmie was in marked contrast with the usual behavior of the monkeys. Selfishness seemed everywhere dominant, while clear indications of sympathetic emotions were rare indeed. The above is undoubtedly the best evidence of anything altruistic that I obtained.

It is possible that Tiny's action was retaliatory, but although it is practically certain that either Gertie or Jimmie inflicted the wound on her finger, I of course cannot be sure that the spirit of revenge stirred her to punish Gertie so severely. Jimmie's part in the whole affair is, however, perfectly intelligible from our human point of view, and there seems no reason to doubt that he did experience something like a feeling of sympathy with his mate, coupled with a feeling of resentment or anger against Tiny.

VI

HISTORICAL AND CRITICAL DISCUSSION OF IDEATIONAL BEHAVIOR IN MONKEYS AND APES

It is my purpose in this section to indicate the relations of my work on monkeys and apes to that of other investigators. Although throughout the report I have used freely the psychological terms idea and ideation, it has been my aim to describe the behavior of my animals rather than to interpret it or speculate concerning its accompaniments. Certain acts I have designated as ideational simply because they seemed to exhibit the essential features of what we call ideational behavior in man. Further study may, and probably will, modify my opinion concerning this matter. It is of prime importance to analyze ideational behavior so that it may be accurately described and satisfactorily defined in terms of its distinguishing characteristics. I had hoped to be able to present a tentative analysis in this report, but the results of my efforts are so unsatisfactory that I do not feel justified in publishing them.

The terms idea and ideation have been used to designate contents of consciousness which are primarily representative. Nowhere have I attempted to indicate different types or grades of ideational behavior and nowhere have I found it necessary to emphasize differences between image and idea. In general, the acts which I have called ideational have been highly adaptive, and the learning processes in connection with which they have appeared have differed strikingly from those of the selective sort in their abruptness of appearance.

Extremely interesting and valuable definitions of ideation and discussions of the characteristics of different sorts of ideas in the light of original observations on monkeys have been presented by Thorndike (1901, pp. 1, 2; 1911, p. 174); Kinnaman (1902, p. 200); and Hobhouse (1915, p. 270). As these authors have contributed importantly to our knowledge of the behavior of monkeys, their discussions of the meaning of terms are especially valuable. Serviceable definitions are to be found, also, in Romanes (1900), Morgan

(1906), Washburn (1908), and Holmes (1911).

Evidences of Ideation in Monkeys Aside from anecdotal and traveller's notes on the behavior of monkeys and apes we have only a scanty literature. In fact, the really excellent articles on the behavior and mental life of these animals may be counted on one's fingers; and not more than half of these are experimental studies. I shall, in this brief historical sketch, neglect entirely the anecdotal literature, since my own work is primarily experimental, and since its results should naturally be compared with those of other experimenters.

Thorndike (1901), the American pioneer in the application of the experimental method to the study of mind in animals, published the first notable paper on the psychology of monkeys. His results force the conclusion that "free ideas" seldom appear in the monkey mind and have a relatively small part in behavior. That the species of Cebus which he observed exhibits various forms of ideation he is willing to admit. But he insists that it is of surprisingly little importance in comparison with what the general behavior of monkeys as known in captivity and as described by the anecdotal writers have led us to expect. It is important to note, however, that Thorndike's observations were limited to Cebus monkeys which, as contrasted with various old world types, are now considered of relatively low intelligence.

In many respects the most thoroughgoing and workmanlike experimental study of monkeys is that of Kinnaman (1902), who has reported on the study of various forms of response in _P. rhesus_. He presents valuable data concerning the learning processes, sensory discrimination, reaction to number, and to tests of imitation. His results indicate a higher level of intelligence than that discovered by Thorndike, but this is almost certainly due to difference in the species observed. Kinnaman goes so far as to say "We have found evidence, also, of general notions and reasoning, both of low order" (p. 211).

The contribution of Hobhouse (1915) to our knowledge of the mental life of monkeys, although in a measure experimental, is based upon relatively few and unsystematic observations as contrasted with those of Thorndike and Kinnaman. It appears, however, that Hobhouse's experiments were admirably planned to test the ideational capacity of his subjects, and one can not find a more stimulating discussion of ideation than that contained in his "Mind in

Evolution." The results of his tests made with a _P. rhesus_ monkey are similar to those of Kinnaman, for almost all of them indicate the presence and importance of ideas.

Watson (1908) in tests of the imitative ability of _P. rhesus_ saw relatively little evidence of other than extremely simple forms of ideation. But in contrast with his results, those obtained by Haggerty (1909), in a much more extended investigation in which several species of monkey were used, obtained more numerous and convincing evidences of ideation in imitative behavior. Although this author wholly avoids the use of psychological terms, seeking to limit himself to a strictly objective presentation of results, it is clear from an unpublished manuscript (thesis for the Doctorate of Philosophy, deposited in the Library of Harvard University) that he would attribute to monkeys simple forms of ideational experience.

Witmer (1910) reports, in confirmation of Haggerty's results, intelligently imitative behavior in _P. irus_.

The work of Shepherd (1910) agrees closely, so far as evidences of ideation are concerned, with that of Thorndike. He obviously strives for conservatism in his statements concerning the adaptive intelligence of his monkeys, all of which belonged to the species _P. rhesus_. At one point he definitely states that they exhibit ideas of a low order, or something which corresponds to them. Satisfactory evidences of reasoning he failed to obtain.

Franz's (1907, 1911) studies of monkeys, unlike those mentioned above, have for their chief motive not the accurate description of various features of behavior but instead knowledge of the functions of various portions of the brain. His results, therefore, although extremely interesting and of obvious value to the comparative psychologist, throw no special light upon the problem of ideation.

The investigation by Hamilton (1911) of reactive tendencies in _P. rhesus_ and irus yielded preeminently important data concerning complex behavior. For the ingenious quadruple-choice method devised by this observer showed that mature monkeys exhibit fairly adequate types of response. As Hamilton's interest centered in behavior, he did not discuss ideation, but this does not prevent the comparison of his data with those of the present report, and the agreement of his findings with my own is obvious.

My work contrasts sharply with that briefly mentioned above in that I applied systematically and over a period of several months an experimental method suited to reveal problem solving ability. Previously, the so-called problem or puzzle-box method had been used as a means of testing for the presence of ideas. For this I substituted the multiple-choice method. One of the chief advantages of this new method is the possibility of obtaining curves of learning for the solution or attempted solution of relational problems of varying difficultness. I am confident that these curves of learning will prove far more valuable than such data as are yielded by the puzzle-box method.

The Pithecus monkeys, which I studied intensively, yielded relatively abundant evidences of ideation, but with Thorndike I must agree that of "free ideas" there is scanty evidence; or rather, I should prefer to say, that although ideas seem to be in play frequently, they are rather concrete and definitely attached than "free." Neither in my sustained multiple-choice experiments nor from my supplementary tests did I obtain convincing indications of reasoning. What Hobhouse has called articulate ideas, I believe to appear infrequently in these animals. But on the whole, I believe that the general conclusions of previous experimental observers have done no injustice to the ideational ability of monkeys. It is clearly important, however, that we always should take into account the species of animal observed, for unquestionably there are extreme differences in mental development among the monkeys.

As I view my results in the light of their relations to earlier work, I am strongly impressed with the importance of the use of improved methods for the study of complex behavior. The delayed reaction method of Hunter, the quadruple-choice method of Hamilton, and my multiple-choice method offer new and promising approaches to forms of activity which thus far have been only superficially observed.

The ability exhibited by Skirrl to try a method out and then to abandon it suddenly is characteristic of animals high in intelligence. Most of the problems which I presented to my animals would be rated as difficult by psychologists, for as a rule they involved definite relations and demanded on the part of the subject both perception of a particular relation and the ability to remember or re-present it on occasion.

I was greatly surprised by the slow progress of the monkeys toward the solution of these problems. It had been my supposition that they would solve them more quickly than any lower type of mammal, but as a matter of fact they succeeded less well than did pigs. Their behavior throughout the work proved that of far greater significance for the experimenter than the solution of a problem is definite knowledge of the modes of behavior exhibited from moment to moment, or day to day. This is true especially of those incidental or accidental modes of response which so frequently appeared in connection with my work that I came to look upon them, the surprises of each day, as my chief means of insight.

Evidences of Ideation in Apes Reliable literature of any sort concerning the behavior and mental life of the anthropoid apes is difficult to find, and still more rare are reports concerning experimental studies of these animals. There are, it is true, a few articles descriptive of tests of mental ability, but even these are scarcely deserving of being classed as satisfactory experimental studies of the psychology of the ape. I have the satisfaction of being able to present in the present report the first systematic experimental study of any feature of the behavior of an anthropoid ape.

Among the most interesting and valuable of the descriptions which may be classed among accounts of tests of mental ability is Hobhouse's (1915) study of the chimpanzee. The subject was an untrained animal, so far as stated, of somewhat unsatisfactory condition because of timidity. Nevertheless, Hobhouse was able to obtain from him numerous and interesting responses to novel situations, some of which may be safely accepted as evidences of ideation of a fairly high order.

Similar in method and result to the work of Hobhouse is that of Haggerty (unpublished thesis for the Doctorate of Philosophy, deposited in the Library of Harvard University). Haggerty's tests of the ability of young orang utans and chimpanzees to solve simple problems and to use tools in various ways yielded results which contrast most strikingly with those obtained in his experimental study of the imitative tendency in monkeys. His observations, had he committed himself to anything approaching interpretation, doubtless would have led him to conclusions concerning the ideational life of these animals very similar to those of Hobhouse.

Koehler, working in the Canary Islands, has, according to information which I have received from him by letter, made certain experiments with orang utans and chimpanzees similar to those of Hobhouse and Haggerty. His results I am unable to report as I have scanty information concerning them. They are, presumably, as yet unpublished.

In his laboratory at Montecito, California, Hamilton has from time to time kept anthropoid apes, but without special effort to investigate their ideational behavior. He has most interesting and valuable data concerning certain habits and instincts, all as yet unpublished.

To a congress of psychologists Pfungst (1912) briefly reported on work with anthropoid apes in certain of the German zoological gardens. His preliminary paper does not enable one to make definite statements concerning either his methods or such results as he may have obtained concerning ideational behavior. So far as I know, he has not as yet published further concerning his investigation.

Merius (1867) has described interesting observations concerning the mental life of the chimpanzee. But this, like all of the work previously mentioned, is rather in the nature of casual testing than thoroughgoing, systematic, and analytic study.

In addition to the above reports, there are a few concerning the behavior of apes which have been especially trained for purposes of exhibition. Most interesting of these is that of Witmer (1909), who studied in exhibitions and in his own laboratory the behavior of the chimpanzee Peter. The varied forms of intelligently adaptive behavior exhibited by this ape convinced Witmer of ideational experience and even of an approach to reasoning. In his brief report he expresses especial interest in the possibility of educating this "genius among apes" to the use of language.

A chimpanzee named Consul was observed several years ago by Hirschlaff (1905), and his tricks were interestingly described from the pedagogical standpoint.

Similar in character is Shepherd's (1915) brief description of the stage behavior of Peter and Consul, both chimpanzees. It is impossible to determine from the account whether these animals are the same as were observed by both Witmer and Hirschlaff. As no reference is made in Shepherd's paper to other descriptions of the behavior of these animals and as he adds nothing to what had already been presented, the reader obtains no additional light on ideation.

I have mentioned only samples of the articles on trained anthropoids. All are necessarily descriptions of the behavior of individuals who had been trained not for psychological purposes but for the vaudeville stage, and although such observations unquestionably have certain value for comparative psychology, it is well known that unless an observer knows the history of an act, he is not able to evaluate it in terms of intelligence and is especially prone to overestimate its value as evidence of ideation.

There remain studies of the apes, dealing primarily with behavior and mental characteristics, which are slightly if at all experimental and deserve to be ranked as naturalistic accounts. Such is, for example, the book of Sokolowski (1908), in which attention is given to the characteristics of young as well as fairly mature specimens of the gorilla, chimpanzee and orang utan.

The various publications of Garner (1892, 1896, 1900) deal especially with the language habits of monkeys and apes, but observations bearing on ideation are reported.

Wallace (1869) describes certain features of the behavior of an infant orang utan whose mother he shot in Borneo. He also reports observations concerning the behavior of adult orang utans, many specimens of which were shot by him during his travels.

Early in the last century, Cuvier (1810) interested himself in studies of the intellectual characteristics of the orang utan, and his data, taken with those of Wallace, Sokolowski, and others similarly interested in the natural history of mind, give one a valuable glimpse of the life of the anthropoid ape.

Finally, the data brought together by Brehm (1864, 1875, 1888) in his famous Tierleben; by Darwin (1859, 1871) in "The Origin of Species," and other works, by Romanes (1900), especially in his books on mental evolution, by C. Lloyd Morgan (1906) in his several works on comparative psychology, and by Holmes (1911) in his discussion of the evolution of intelligence, contribute not unimportantly to our all too meagre knowledge of the mental life of the anthropoid apes.

My own results, viewed in the light of what one may learn from the literature, stand out as unique because of the method of research. Never before, so far as I have been able to learn, has any ape been subjected to observation under systematically controlled conditions for so long a period as six months. Moreover, my multiple-choice method has the merit of having yielded the first curve of learning for an anthropoid ape. This fact is especially interesting when one considers the nature of the particular curve. For so far as one may say by comparing it with the curves for various learning processes exhibited by other mammals, it is indicative of ideation of a high order, and possibly of reasoning. I do not wish to exaggerate the importance of my results, for as contrasted with what might be obtained by further study, and with what must be obtained if we are adequately to describe the mind of the orang utan, they are meager indeed.

Especially noteworthy, as evidences of ideation, in the results yielded by the multiple-choice method are (1) the use by the orang utan of several different methods in connection with each problem; (2) the suddenness of transition from method to method; (3) the final and perfect solution of problem I without diminution of the initial errors; (4) the dissociation of the act of turning in a circle from that of standing in front of a particular box.

To these features of behavior others of minor importance might be added. But as they have been sufficiently emphasized in the foregoing detailed descriptions, I need only repeat my conclusion, from the summation of evidence, that this young orang utan exhibited numerous free ideas and simple thought processes in connection with the multiple-choice experiment. His ultimate failure to solve the second problem is peculiarly interesting, although in the light of other features of his behavior by no means indicative of inferior intelligence.

The various supplementary experimental tests which I employed are in no wise importantly distinguished from those used by other observers. The box stacking experiment has, according to my private information, been used by Koehler. It is obviously important that such tests be applied in the same manner to individuals not only of the different genera of anthropoid apes, but of different ages, sex, and condition of training.

The box stacking experiment, although it yielded complete success only as a result of suggestion on my part, proved far more interesting during its progress than any other portion of my work. In connection with it, the orang utan exhibited surprisingly diverse and numerous efforts to meet the demands of the situation. It is fair to characterize him as inventive, for of the several possible ways of obtaining the banana which were evident to the experimenter, the ape voluntarily used all but two or three, and one of these he subsequently used on the basis of imitation.

Had Julius been physically and mentally mature, my results would undoubtedly have been much more impressively indicative of ideas, but even as matters stand, the survey of my experimental records and supplementary notes force me to conclude that as contrasted with the monkeys and other mammals, the orang utan is capable of expressing free ideas in considerable number and of using them in ways highly indicative of thought processes, possibly even of the rational order. But contrasted with that of man the ideational life of the orang utan seems poverty stricken. Certainly in this respect Julius was not above the level of the normal three-year-old child.

In common with other observers, I have had the experience of being profoundly impressed by the versatility of the ape, and however much I might desire to disprove the presence of free ideas and simple reasoning processes in the orang utan, I should feel bound to accept many of the results of my tests as evidences of such experience.

I have attempted to indicate briefly the historical setting of my investigation. I propose, now, in the concluding section, to look forward from this initial research and to indicate as well as I may in a few words the possibilities of results important for mankind from the thorough study of the monkeys and anthropoid apes.

VII

PROVISION FOR THE STUDY OF THE PRIMATES, AND ESPECIALLY THE MONKEYS AND ANTHROPOID APES[1]

[Footnote 1: Much of the material of this section was published originally in Science (Yerkes, 1916).]

I should neglect an important duty as well as waste an opportunity if in this report I did not call attention to the status of our knowledge concerning the monkeys and apes and present the urgent need of adequate provision for the comparative study of all of the primates.

Although for centuries students of nature have been keenly interested in the various primates, the information which has been accumulated is fragmentary and wholly inadequate for generally recognized scientific and practical needs. There is a voluminous literature on many aspects of the organization and lives of the monkeys and apes, but when one searches in it for reasonably connected and complete descriptions of the organisms from any biological angle, one, is certain to meet disappointment.

Concerning their external characteristics we know much; and our classifications, if not satisfactory to all, are at least eminently useful. But when one turns to the morphological sciences of anatomy, histology, embryology, and pathology, one discovers great gaps, where knowledge might reasonably be expected. Even gross anatomy has much to gain from the careful, systematic examination of these organisms. With still greater force this statement applies to the studies of finer structural relations. Little is known concerning the embryological development and life history of certain of the primates, and almost nothing concerning their pathological anatomy.

Clearly less satisfactory than our knowledge of structure is the status of information concerning those functional processes which are the special concern of physiology and pathology. Certain important experimental studies have been made on the nervous system, but rarely indeed have physiologists dealt systematically with the functions of other systems of organs. There are almost no satisfactory physiological descriptions of the monkeys, anthropoid apes, or lower primates.

SUB-DIVISIONS OF THE ORDER PRIMATES

Order _Sub-orders_ Families ,- a. PROSIMII (Lemurs and Aye-Ayes) | | ,- i. Hapalidae (Marmosets) | | ii. Cebidae (Howling Monkeys, PRIMATES -+ | Tee Tees, Squirrel Monkeys, | | Spider Monkeys, and Capuchin | | Monkeys) `- b. ANTHROPOIDEA ... -+ iii. Cercopithecidae (Baboons | and Macaques) | iv. Simiidae (Gibbons, Orangs, | Chimpanzees, and Gorillas) `- v. Hominidae (Man)

When we turn to the science of genetics we meet a similar condition, for the literature reveals only scattered bits of information concerning heredity in the primates. No important experimental studies along genetic lines have been made with them, and such general observations from nature as are on record are of extremely uncertain value. Were one to insist that we know nothing certainly concerning the relation of heredity in other primates than man, the statement could not well be disputed.

Occasionally in recent years students of human diseases have employed monkeys or apes for experimental tests, but aside from the isolated results thus obtained, extremely little is known concerning the diseases peculiar to the various types of infra-human primates or the significant relations of their diseases to those of man.

Next in order of extent to our morphological knowledge of these organisms is that of their behavior, mental life, and social relations. But certainly no one who is conversant with the behavioristic, psychological and sociological literature could do otherwise than emphasize its incompleteness and inadequacy. For our knowledge of behavior has come mostly from naturalistic observation, scarcely at all from experimentation; our knowledge of social relations is obviously meager and of uncertain value; and finally, our knowledge of mind is barely more than a collection of carelessly drawn inferences.

This picture of the status of scientific work on the primates, although not overdrawn, will doubtless surprise many readers, and even the biologist may find himself wondering why we are so ignorant concerning the lives of the organisms most nearly akin to us, and naturally of deepest interest to us. The reasons are not far to seek. Most scientific investigators are forced by circumstances to work with organisms which are readily obtained and easily kept. The primates have neither of these advantages, for many, if not most of them, are expensive to get and either difficult or expensive to keep in good condition. Clearly, then, our ignorance is due not to lack of appreciation of the scientific value of primate research but instead to its difficultness and costliness.

Strangely enough, the practical importance of knowledge of the primates has seldom been dwelt upon even by those biologists who are especially interested in it. It is, therefore, appropriate to emphasize the strictly human value of the work for which I am seeking provision.

During the past few years it has been abundantly and convincingly demonstrated that knowledge of other organisms may aid directly in the solution of many of the problems of experimental medicine, of physiology, genetics, psychology, sociology, and economics. In the light of these results, it is obviously desirable that all studies of infrahuman organisms, but especially those of the various primates, should be made to contribute to the solution of our human problems.

To me it seems that thoroughgoing knowledge of the lives of the infrahuman primates would inevitably make for human betterment. Through the science of genetics, as advanced by experimental studies of the monkeys and anthropoid apes, practical eugenic procedures should be more safely based and our ability to predict organic phenomena greatly increased. Similarly, intensive knowledge of the diseases of the other primates in their relations to human diseases should contribute importantly to human welfare. And finally, our careful studies of the fundamental instincts, forms of habit formation, and social relations in the monkeys and apes should lead to radical improvements in our educational methods as well as in other forms of social service.

Along theoretical lines, no less than practical, systematic research with the primates should rapidly justify itself, for upon its results must rest the most significant historical or genetic biological descriptions. It is beyond doubt that genetic psychology can best be advanced to-day by such work, and what is obviously true of this science is not less true of all the biological sciences which take account of the developmental or genetic relations of their events.

In view of the probable values of increasingly complete accounts of primate life, it seems far from extravagant to insist that the securing of adequate provision for systematic and long continued research is the most important task for our generation of biologists and the one which we shall be least excusable for neglecting. Indeed, when one stops to reflect concerning the situation, it seems almost incredible that the task has not been accomplished.

Some ten years ago Professor John B. Watson (1906) entered a plea for the founding of a station for the experimental study of behavior. He made no special mention of work with the monkeys and apes, but it is clear from the problems which he enumerates that he would consider them most important subjects for observation. Professor Watson's plea has apparently been forgotten by American biologists, and it seems not inappropriate to revive it at this time. For surely we have advanced sufficiently along material and scientific lines during the last ten years to render possible the realization of his hope.

To my knowledge, only one definite attempt has thus far been made to gain special provision for the study of the primates. Somewhere about the year 1912 there was established on Tenerife, one of the Canary Islands, a modest station for the study of the anthropoid apes. I have already referred to it briefly on page 1. The plan and purpose of this station, which is of German origin, have been presented briefly by Rothmann (1912). From personal communications I know that a single investigator has been in residence at the station since its founding and that psychological and physiological results of value have been obtained, but no published reports have come to my attention.

When I first heard of the existence of the German anthropoid station I naturally thought of the possibility of cooperative work, but the events of the past two years have rendered the chances of cooperation so remote that it now seems wholly desirable and indeed imperative to seek the establishment of an American station, which, unlike the German station, shall provide adequately not only for the study of the anthropoid apes but for that of all of the lower primates. It should be the function of such a station or research institute (1) to bring together and correlate all the information at present available; (2) to fill in existing gaps observationally and thus complete and perfect our knowledge of these organisms; (3) to seek to bring all available information to bear upon the problems of human life.

Hitherto the unsatisfactoriness of progress has been due to the lack of a definite plan and program. Every investigator has gone his own way, doing what little his personal means and opportunity rendered possible. The time has at last come when concerted action seems feasible as well as eminently desirable. I am therefore offering a plan and program which, if wisely developed, should lead ultimately to fairly complete and practically invaluable knowledge of the lives of all of the primates. There should be provided in a suitable locality a station or research institute which should offer adequate facilities (1) for the maintenance of various types of primate in normal, healthy condition; (2) for the successful breeding and rearing of the animals, generation after generation; (3) for systematic and continuous observation under reasonably natural conditions; (4) for experimental investigations from every significant biological point of view; (5) for profitable cooperation with existing biological institutes or departments of research throughout the world.

The station should be located in a region whose climate is highly favorable to the life of many of the lower primates as well as to that of man. Such a location is by no means easy to find. Because of my intense interest in the subject, I have, during the past five years, prospected in various parts of the world for a satisfactory site. I shall now attempt to indicate the chief requirements and also the foremost advantages and disadvantages of several regions which have been considered. It is first of all requisite that the climate be such as to agree with the organisms to be studied and such, also, as to render their breeding normal and dependable. Second in importance is its satisfactoriness for the life and scientific productiveness of the observer. While certain tropical localities would meet the first requirement perfectly, they would prove extremely unsatisfactory for research activity. It therefore seems essential to find a region whose climate shall reasonably meet the needs of the experimenter while adequately meeting those of the animals to be studied.

A further factor which has important bearing upon the productiveness of the observer is the degree of isolation from civilization and from other scientific work. No scientist can long work effectively, even in a reasonably healthy and stimulating climate, if entirely cut off from similar interests and activities. It is therefore desirable, if at all possible, to discover a location in the midst of civilization and with reasonably good opportunities for scientific associations.

With these several desiderata before us, I shall call attention to a number of possible sites for a station, several of which I have visited. Southern California, and especially the portion of the State between Santa Barbara and San Diego, promises fairly well. It is definitely known that certain, if not all, species of monkey will breed there fairly satisfactorily, and although it has not yet been demonstrated, there is no reason to suppose that in certain regions the anthropoid apes might not also be kept in perfect health and successfully bred. The main advantages of this general region are (a) a climate which promises to be reasonably satisfactory for many if not all of the primates; (b) admirable climatic conditions for investigators; (c) wholly satisfactory scientific and cultural environment for the staff of a station. The most significant disadvantages are (a) a temperature, which is at times a trifle too low for the comfort of certain of the monkeys and apes. It is by no means certain, however, that they would not usually adapt themselves to it. (b) The necessity of importing all of the animals and of having to rely upon successful acclimatization. Of course it is to be assumed that importation would be necessary only at the outset of such work, since the animals later should replenish themselves within the confines of the station.

Florida offers possibilities somewhat similar to those of southern California, but as I have not had opportunity to examine the conditions myself, I can say only that in view of such information as is available the advantage seems to be greatly in favor of the latter.

Cuba, Jamaica, Porto Rico, and for that matter, several of the West Indies, offer possible sites for a successful station. I have reasonably intimate personal knowledge only of the conditions in Jamaica. The major advantages in the West Indies are (a) suitable climatic conditions and food supply for the animals; and (b) reasonably satisfactory climatic conditions for the staff. These are, however, more than counterbalanced in my opinion by the following serious disadvantages: (a) the relative isolation of the investigators from their fellow scientists; (b) the necessity of importing all of the animals originally used; (c) the risk of destruction of the station by storms.

It is definitely known that anthropoid apes as well as monkeys can be successfully kept, bred, and reared in the West Indies. During the past year, on the estate of Doha Rosalia Abreu, near Havana, Cuba, a chimpanzee was born in captivity. A valuable account of this important event and of the young ape has been published by Doctor Louis Montan?(1915). It therefore seems practically certain that regions could be found readily on Jamaica, Porto Rico, or smaller islands, which would be eminently satisfactory for the breeding of apes.

There are obvious reasons why an American station for the study of the primates should be located on territory controlled by the United States Government, and if a tropical location proves necessary, it would probably be difficult to find more satisfactory regions, aside from the inconveniences and risk of importation and the relative isolation of the investigators, than are available on Porto Rico.

I have not seriously considered the possibility of locating an American station on the continent of Africa, for although two of the most interesting and important of the anthropoid apes, the gorilla and the chimpanzee, are African forms, while many species of monkey are either found there or could readily be imported, it has seemed to me that the islands of the West and East Indies and the portions of the United States referred to above are much to be preferred over anything available in Africa.

In the East, Borneo, the Philippine Islands, and Hawaii are well worth considering. Borneo is the home of the gibbon and of at least one species of orang utan, and in addition to these important assets, it presents the advantages of (a) a wholly suitable climate and food supply for monkeys and apes; and (b) climatic conditions for investigators which, I am informed by scientific friends, are nearly ideal. For investigators the most serious disadvantage here, as in all other parts of the East, would be the isolation from other scientific work and workers.

The possibilities of Central America I considered several years ago when it seemed to me possible that work might profitably be done with monkeys and apes on the Canal Zone. The advantages are (a) a climate which promises fairly well for the animals; and (b) reasonable accessibility from the United States. The disadvantages are (a) a far from ideal climate for long continued scientific work; and (b) an environment which from the cultural and scientific point of view leaves much to be desired.

Were a permanent psycho-biological station for the study of the primates to be established in southern California, it would, even though wholly satisfactory conditions for the breeding, rearing, and studying of the animals were maintained, furnish more or less inadequate opportunity for the observation of the animals under free, natural conditions. It would therefore be necessary, to supplement the work of such a station by field work in Borneo, Sumatra, Africa, India, South America, and such other regions as the species of organism under consideration happen to inhabit.

Considering equally the needs of the experimenter and the demands of the animals, it seems to me reasonable to conclude that southern California should be definitely proved unsuitable before a more distant site were selected. For the information which I have been able to accumulate convinces me that it would in all probability be possible successfully to breed and keep the primates there, and it is perfectly clear that in such event the output of a station would be enormously greater because of the more favorable conditions for research than in any tropical region or in a more isolated location.

Assuming that satisfactory provision in the shape of a scientific establishment for the study of the primates in their relations to man were available, the following program might be followed: (1) Systematic and continuous studies of important forms of individual behavior, of social relations, and of mind; (2) experimental studies of physiological processes, normal and pathological, and especially of the diseases of the lower primates, in their relations to those of man; (3) studies of heredity, embryology, and life history; (4) research in comparative anatomy, including gross anatomy, histology, neurology, and pathological anatomy.

Each of these several kinds of research should be in progress almost continuously in order that no materials or opportunities for observation be needlessly wasted. Because of the nature of the work, it would be necessary to provide, first of all, for those functional studies which demand healthy and normally active organisms, whose life history is intimately and completely known. This is true of all studies in behavior, whether physiological, psychological, or sociological. Simultaneously with behavioristic observations and often upon the same individuals, genetic experiments might be conducted. This would be extremely desirable because of the relatively long periods between generations. After the usefulness of an animal in behavioristic or genetic inquiries had been exhausted, it might be made to render still further service to science in various experimental physiological, or medical inquiries. And finally, the same individual might ultimately be used for various forms of anatomical research. Thus, it is clear that the scientific usefulness of a lemur, a monkey, or an ape might be maintained at a high level throughout and even beyond the period of its life history.

The program thus briefly sketched would provide either directly or indirectly for work on every aspect of primate life. Especially important would be the intimacy of interest and cooperation among investigators, for the comparative method should be applied consistently and to the limit of its value. The results of various kinds of observation should be correlated so that there should ultimately emerge a unitary and practically valuable account of primate life, to replace the patchwork of information which we now possess.

Because of the costliness of maintaining and breeding the monkeys and apes, it is especially desirable that the several kinds of research mentioned above should be conducted. Indeed, it would seem inexcusably wasteful to attempt to maintain a primate or anthropoid station for psychological observations alone, or for any other narrowly limited biological inquiry.

Furthermore, the station should be permanent, since for many kinds of work it would be essential to have intimate knowledge of the life history and descent of an individual. With the lower primates, a generation might be obtained in from two to five years; with the higher, not more frequently, probably, than from ten to fifteen years. It therefore seems not improbable that the value of the work done in such a station would continue to increase for many years and would not reach its maximum short of fifty or even one hundred years.

A staff of several highly trained and experienced biologists would be needed. The following organization is suggested as desirable, although, as indicated below, not necessarily essential in the beginning: (1) An expert especially interested in the problems of behavior, psychology, and sociology, with keen appreciation of practical as well as of theoretical problems; (2) an assistant trained especially in comparative physiology; (3) an expert in genetics and experimental zoology; (4) an assistant with training and interests in comparative anatomy, histology, and embryology; (5) an expert in experimental medicine, who could conduct and direct studies of the diseases of man as well as of the lower primates and of measures for their control; (6) an assistant trained especially in pathology and neurology.

To this scientific staff of six highly trained individuals there should be added a business manager, a clerical force of three individuals, a skilled mechanician, a carpenter, and at least four laborers.

The annual expenditures of an institute with such a working staff, would in southern California, approximate fifty thousand dollars. It would therefore be necessary that it have an endowment of approximately one million dollars.

In the absence of this foundation it would, of course, be possible to make a reasonably satisfactory beginning on the work which has been outlined in the following less expensive manner. A working plant might be established, on ground rented or purchased at a low figure, for about ten thousand dollars; the salary of a director, assistants, a clerical helper, and combined mechanic and laborer might be estimated at the same figure; the cost of animals and of maintenance of the plant would approximate five thousand dollars. Thus, we should obtain as an estimate of the expenditures for the first year twenty-five thousand dollars. Without expansion, the work might be conducted during the second year for fifteen thousand dollars, and subsequently it might be curtailed or expanded, resources permitting, according as results achieved and in prospect justified.

An institute established on such a modest basis as this still might render largely important scientific service through its own research and through organized cooperation with other existing research establishments. Thus, for example, supposing that behavioristic, psychological, sociological, and genetic inquiries were conducted in the institute itself, animals might be supplied on a mutually satisfactory basis to institutes for experimental medicine, for physiological research, and for anatomical studies. Under such conditions, it is conceivable that extremely economical and good use might be made of all the available primate materials. But it is not improbable that even cooperative research would prove on the whole more profitable, except possibly in the case of morphological work, if investigators could conduct their studies in the institute itself rather than in distant laboratories. In any event, the idea of cooperation should be prominent in connection with the organization of a research station for the study of the primates. For thus, evidently, scientific achievement in connection with these important types of animal might be vastly increased over what would be possible in a single relatively small institution with a limited and necessarily specialized staff of workers.

Despite the fact that biologists generally recognize the importance of the work under consideration and are eager to have it done, it is perfectly certain that we shall accomplish nothing unless we devote ourselves confidently, determinedly and unitedly, with faith, vision, and enthusiasm, to the realization of a definite plan. Our vision is clear,--if we are to gather and place at the service of mankind adequate comparative knowledge of the life of the primates and if we are to make this possible harvest of scientific results count for human betterment, we must bend all our efforts to the establishment of a station or institute for research.

###